The Logic of Collective Choice

THE LOGIC OF COLLECTIVE CHOICE

Thomas Schwartz

Columbia University Press
New York 1986

Library of Congress Cataloging in Publication Data

Schwartz, Thomas.
The logic of collective choice.

Bibliography: p.
Includes index.
1. Social choice. I. Title.
HB846.8.S39 1985 302′.13 85-473
ISBN 0-231-05896-9

Columbia University Press
New York Guilford, Surrey

Printed in the United States of America

Clothbound editions of Columbia University Press books are Smyth-sewn.

For Ellie, Mary Ann, and Bobby

CONTENTS

FOREWORD

Charles R. Plott

Books and articles on social choice theory can have many different purposes. First, the theory has deep roots in social philosophy and concepts of social welfare. If the philosophical interpretation is applied, then the choice function defines the set that society should choose in some moral, ethical, or normative sense. This perspective is probably what Arrow had in mind. A second purpose turns on the fact that the theory has roots in behavioral modeling and game theory. If the behavioral interpretation is applied, then the social choice itself can be interpreted as a game-theoretic solution such as the core. The choice is a prediction about the alternatives that will emerge as a result of the implicit conflicts resolved within a postulated institutional framework. The theory also has an internal logic of its own. Sometimes the purpose of an investigation is no more than an exploration of the logical structure itself to see how its various features are connected. All three of these objectives can be found throughout Schwartz's book.

The first two parts of the book appear to me to be primarily normative although some purely logical excursions can be detected. Schwartz has two very important points to make. First, the use of preference intensities and related concepts of cardinality do not relieve one from the discipline of the theory and the omnipresent impossibility results. A retreat from impossibility theorems to cardinal and measurable utilities does little to avoid paradoxes. The second point is that the source of the negative impossibility results lies in the social rationality postulates. Schwartz thinks that as a consequence of the theory concepts of social preference and social rationality should be removed from social philosophy. The title of his chapter 5 carries the full message.

The third and fourth parts of the book are about behavior. The theory is connected to the theory of coopative games and to institutions. In many respects this is the most amazing part of the book because it exposes the versatility of the theory. Not only can old solution concepts be represented; the chapters are filled with new concepts. Furthermore, concepts of power, multistage decision processes and decentralization can be easily represented in a formal framework. Contrary to what is often believed, the theory is not only about voting.

The book has many new ideas. Other ideas are pulled from a wide-ranging literature. Anyone with a serious interest in social philosophy and game theory will benefit from studying this book carefully. Many students think they understand social choice theory after having worked through the mathematics of an impossibility result. Schwartz's work demonstrates that the implications of the theory go far beyond the technical aspects and that many important and interesting questions remain open.

PREFACE

This book issued from my reflections on the Marquis de Condorcet's Paradox of Voting and Kenneth Arrow's Impossibility Theorem. Its germ was my *Notes on the Abstract Theory of Collective Choice*, versions of which were reproduced between 1970 and 1976 for my students at Stanford and Carnegie-Mellon. I wrote an early version of about two-thirds of the book in 1977 and a penultimate draft in 1980. Meanwhile I published some of its findings in articles, but much of it has not previously appeared in print. Although this is primarily a research monograph rather than a treatise or textbook, its subject matter is broad and foundational, its treatment is self-contained, and its contents are expressly related to the surrounding literature.

The pioneer work of Arrow and Duncan Black spawned my own research, and my debt to them is incalculable. The following scholars have influenced this book through personal communications as well as through their writings: Martin Bailey, Peter Bernholz, Georges Bordes, Gary Cox, James Enelow, John Fellingham, John Ferejohn, Peter Fishburn, Norman Frohlich, William Galston, Allan Gibbard, Bernard Grofman, Thomas Hammond, Melvin Hinich, Jerry Kelley, Mathew McCubbins, Richard McKelvey, Nicholas Miller, Paul Newman, Joe Oppenheimer, Peter Ordeshook, Dennis Packard, Charles Plott, Nicholas Rescher, Jeffrey Richelson, William Riker, Norman Schofield, Amartya Sen, Kenneth Shepsle, Joseph Sneed, Gordon Tullock, Harrison Wagner, Barry Weingast, and everyone whom I have inadvertently neglected to include in this list. Plott and Richelson have been especially influential. The final draft profited from the detailed comments of Jerry Kelley, who combined mastery of subject with the eye and judgment of a copy editor, and Thomas Hammond, who found gaps and obscurities by insightfully modeling my audience.

I thank the University Research Institute of the University of Texas for two major research grants and several minor ones, the Sid Richardson Foundation for a grant through the Institute for Constructive Capitalism of the University of Texas, and Bernard Grofman and the School of Social Science of the University of California, Irvine, for providing the environment and secretarial assistance during the spring of 1980 that made it possible for me to write this book. That institution's peerless word-processing staff produced the penultimate draft, into which final revisions were flawlessly woven by Betty McEuen of the Department of Government, University of Texas. Bernard Gronert, my editor at Columbia, showed a remarkable grasp of disciplinary and subdisciplinary peculiarities and nuances. During the twelve years in which I have worked toward this book, my wife, Ellie, provided understanding, support, and an environment conducive to my efforts.

The Logic of Collective Choice

INTRODUCTION

Within clubs, committees, parliaments, polities, and other corporate creatures, members often make choices collectively rather than individually: they vote, or otherwise express their will, and a choice attributable to all of them but to no one of them is determined thereby. Imagine a state or other organization devoid of any collective-choice process, however informal, infrequent, or exclusive its use. Although possessing governmental functions, such an organization would lack anything ordinarily described as *politics*, or *political activity*. Let an oriental despot's courtiers, counselors, or concubines start engaging in politics, broadly conceived, let them start organizing or competing to affect some governmental actions, and you have the informal, rudimentary beginnings of a collective-choice process—a process in which each participant expresses his will and a choice is thereby made by all of them but no one of them. Political democracy is distinguishable, perhaps, by formal, frequent, popular participation in collective-choice processes. But the mere existence of a collective-choice process is less the mark of political democracy specifically than of politics.

This is an abstract, deductive study of some salient types of collective-choice processes. It is deductive in the sense that my conclusions are theorems; my arguments, proofs. It is comparatively abstract in the following sense: I represent collective-choice processes by mathematical objects that depict relatively little of those processes' structural complexity; this enhances the generality of my conclusions. By and large, those conclusions spell out the precise way each of various broad types of collective-choice process constrains the outcomes that can issue from processes of that type. I think you will find many of these results surprising.

A thematic thread ties them together: When one demands of collective choice the kind of formal "rationality" economists and others have customarily attributed to individual choice, one demands (roughly speaking) that collective choices *maximize* something—or, what comes to much the same thing, one demands that collective choices have a certain *stable* or *equilibrium* or *undominated* character. Apparently one also demands too much. For the demand conflicts, in various circumstances, with other, apparently reasonable and realistic requirements. Yet even theorists who concede that the traditional conception of collective rationality is overly restrictive often remain prisoners of this conception. While not insisting that all collective choices are or should be rational in the traditional sense, some of these theorists still find collective *ir*rationality, as traditionally conceived, to be disturbing, puzzling, even paradoxical, and many still make it their main business to ask in what cases or with what likelihood the traditional conception is applicable, throwing up their hands in bewilderment at the residue cases. My approach is rather to broaden the traditional conception, seeking assumptions that are mathematically fruitful but no longer overly restrictive. These generalized assumptions are motivated by an analysis of institutional and other procedures for arriving at choices rather than by academic tradition, aesthetic appeal, or normative dogma.

Although the results to follow are mathematical theorems, not empirical generalizations or value judgments, I want you to find them useful in understanding and evaluating real political practices and events.

For readers already familiar with collective-choice theory, here is a brief preview:

Part I (chapters 1 and 2), "The Formal Representation of Choice Processes," explains some terms, symbols, and assumptions that play recurring roles throughout later chapters. In much of this book, collective-choice processes are represented by choice functions. Chapter 1 introduces the concept of a choice function along with some related notions, including preference, indifference, ordering relations, and rationality—or "rationality."

Chapter 2 introduces the representation of collective choice as a function of individual preferences. The main topics discussed are domain conditions, cardinal utility, the Independence of Irrelevant Alternatives, and Simple Majority Rule. My domain conditions, later used in impossibility theorems, are weaker in important respects than comparable conditions found in the literature. For example, mine do not require the domain of admissible preference profiles to contain any profiles of *nonlinear* individual orderings. This is not to say that they *restrict* the domain to profiles of linear orderings, only that they are *compatible* with such a restriction. What makes this important is that extant preferential-voting systems often require voters to produce linear rankings on their ballots. I present the Independence of Irrelevant Alternatives as an informational restriction. Insofar as collective choices reflect individual preferences, the Independence condition requires that collective choices reflect only such preferential information as could be got by taking a vote. One might carry this idea a step further by requiring that collective choices reflect only such *information* as could be got by taking a vote, hence that collective choices be independent of all nonpreferential factors. This requirement provides a rationale for May's set of necessary and sufficient conditions for Simple Majority Rule.

Part II (chapters 3–5), "Collective Choice and Collective Rationality," is about impossibility theorems and their significance—the paradoxes of collective choice and their solution. I begin with the Classical Voting Paradox—Condorcet's three-man, three-alternative example—and some simple generalizations of it before turning to Arrow's Theorem. One of the conditions involved in Arrow's Theorem is Collective P + I-Transitivity (my name for it), which says that collective preference and collective indifference are both transitive. This is stronger than Collective P-Transitivity (transitivity of collective strict preference), which in turn is stronger than Collective P-Acyclicity. Yet of these "rationality" conditions, only the last is involved in the Classical Voting Paradox. Thus, Arrow's Theorem does not generalize the Classical Voting Paradox.

Seeking an Arrow-style theorem that does generalize the Classical Voting Paradox, I prove a series of impossibility theorems that rely on progressively weaker transitivity-related conditions. Besides an impossibility theorem based on the well-known oligarchy result for Collective P-Transitivity, I prove that Arrow's original theorem holds for *any* transitivity condition stronger than Collective P-Transitivity but weaker than Collective P + I-Transitivity so long as the domain condition is suitably strengthened. I then prove three impossibility theorems based on Collective P-Acyclicity and weakened versions thereof. The first comes from my 1970 paper on the subject, the first publication of an Arrow-style impossibility theorem based on Collective P-Acyclicity—or, for that matter, any weakening of Collective P + I-Transitivity. The crucial addition to Arrow's conditions is that no two alternatives are collectively indifferent (tied) when one is preferred to the other by all but a single individual. The second theorem is similar to that of Mas Colell and Sonnenschein (1972), who assumed Positive Responsiveness, but my weaker domain condition requires an entirely different proof. The third theorem, which comes from a 1982 paper of mine, generalizes the second in two chief ways: first, Positive Responsiveness, which says that any single individual can break any tie, is replaced by the less stringent condition that any coalition comprising a fifth of society can break any tie; second, Collective P-Acyclicity is replaced by the condition that, for some possible cycle size, there is no collective-preference cycle of exactly that size. This means that the remaining conditions imply the existence, not just of a cycle (corresponding to some preference profile), but of a cycle of any given size, including one that exhausts the given set of alternatives.

I next explore the extent of the phenomenon revealed by these impossibility theorems, examining domain restrictions, the role of the Independence condition, the effect of allowing interpersonal utility comparisons, and the analogous problem of individual multidimensional choice. After a brief survey of Single Peakedness, stability in spatial models, and similar material, I introduce a domain restriction, called Condorcet Freedom, which is more general than Single Peakedness, Value Restriction, and others of their ilk. I prove that this condition guarantees Collective P-Transitivity, not only under Simple Majority Rule, but under a large category of voting rules, including the special-majority rules. My treatment of Independence conditions, interpersonal comparisons, and individual multidimensional choice is based on my 1970 paper, whose impossibility theorem uses a weakened Independence condition, allows substantial reliance on interpersonal utility comparisons, and can be interpreted in terms of multidimensional choice of many types.

Part II concludes with an argument: It is the "rationality" conditions—Collective P + I-Transitivity, Collective P-Acyclicity, and the like—that are responsible for the paradoxes of collective choice. These conditions cannot be justified in the way the other conditions can, and arguments in their behalf collapse under close scrutiny.

Part III (chapters 6–8), "The General Theory of Solutions," is about choice-set or solution specifications. One such specification is the Binary-Choice Property (BICH), the core of the traditional "rationality" assumptions, according to which the choice set from a given potential feasible set is the set of elements undominated under the relation *P* of pairwise choice. This requires *P* to be acyclic, a drawback avoided by three other choice-set specifications: the Generalized Optimal Choice Axiom (GOCHA), the Generalized Top Choice Assumption (GETCHA), and the Solution Condition (SOCO).

The GOCHA set (the choice set from a given set specified by the GOCHA condition) is the union of minimum *P*-undominated subsets (no feasible alternative outside such a subset bears *P* to any alternative inside it). It is also the "top cycle" set. The GETCHA set is the minimum *P*-dominant subset (every alternative in such a set bears *P* to every feasible alternative outside it). By interchanging *P* (strict preference) and *R* (preference or indifference), we can define either set the same way the other is defined. Both GOCHA and GETCHA can be cast in terms of undominated subsets, dominant subsets, top cycles, and transitive closures, and both enjoy interesting axiomatic characterizations. Both also specify sets that tend to be large and can contain Pareto-inefficient members. Although this impairs their predictive ability and their normative appeal, we will find in chapters 9 and 10 that the GOCHA and GETCHA sets comprise the precise output of certain important types of institutional processes.

Unlike BICH, GOCHA, and GETCHA, SOCO does not specify choice sets (or a choice function) solely in terms of the relation *P* of collective preference (pairwise choice). SOCO is couched in terms of individual preferences and "power structures." Applied to a majoritarian power structure, SOCO specifies a choice set that is included in the GOCHA, GETCHA, and Pareto sets. Thus, it yields a voting rule that behaves well in comparison with others on the market. But its main value as a choice-set or solution specification does not emerge until chapter 11.

In Part IV (chs. 9–11), "Collective-Choice Processes Dissected," I shift attention from theorists' conundrums to the institutional structures and procedures through which collective choices actually are made. What effects do these have on collective choices? I begin with *multistage* choice processes, in which the choice from a given set is made, not in one fell swoop, but in a series of choices from various subsets. One might, for example, choose from a set of four alternatives by choosing between two of them, then choosing between the alternative thus chosen and a third alternative, and finally pitting the victor in the latter contest against the remaining alternative. Or one might choose from one set by dividing it into two subsets and choosing from each subset, then choosing between the two chosen alternatives. With the feasible set and the underlying choice function held constant, a multistage choice process can choose any point in the GETCHA set, depending on (1) the rule for breaking ties and (2) the pattern by which the choice from the whole set is decomposed into choices from subsets. This remains true when factors (1) and (2) are severely constrained in natural ways. Because the GETCHA set is so large, this means that multistage choice processes are quite indeterministic—or rather that they depend to a great extent on seemingly arbitrary, hard-to-predict factors.

In chapter 10, I examine one of the most important yet neglected parts of the total process by which any choice is made, namely, the identification or selection of the feasible set—the choice of the set from which to choose. Often any one of a number of sets can, with equal justification, be treated as *the* feasible set. How sensitive are familiar choice processes to the selection of a feasible set? I examine a range of progressively weaker *insensitivity conditions*—conditions which say of a given choice function that it is insensitive to certain changes in the feasible set. These conditions are all variations of the Weak Axiom of Revealed Preference (WARP). Each is equivalent to one of the rationality conditions earlier found to be overly restrictive.

I also examine three processes by which a feasible set might be selected. One consists in starting with a large set and winnowing clearly rejectable alternatives. Another consists in making a provisional choice and successively revising it in the face of newly identified feasible alternatives. This is equivalent to a multistage process. The third consists in making a choice from a small set of clearly exhaustive but comparatively inspecific alternatives (war vs. peace, say), then making this choice ever more specific (nuclear vs. conventional war) as the need arises. Depending on the precise way in which maximally specific alternatives are grouped together to form less specific alternatives, this process can yield any choice in the GOCHA set.

Chapter 11 introduces the analysis of alternatives into *issues*: every alternative is an ordered m-tuple (say) comprising one position on each of m issues. This enables us to talk about decentralization, cooperation, exchange (of goods, votes, or whatnot), the effects of combining versus separating issues, and individual compliance with collective choices. Among other things, I prove the universal instability theorem of my 1981 paper by that name. It says that if any exchange of support across issues (including any packaging of positions) is essential to choosing a given outcome, then that outcome has to be unstable: some group must have the power and incentive to overturn it. This is true for any power structure (however complex or unmajoritarian), any feasible set (however oddly or severely constrained), and any set of individual preferences (however exotic or perverse). Cooperation problems (as in the Prisoners' Dilemma) and Sen's celebrated Liberal Paradox are instances of this theorem: there is no special collective-choice problem of liberalism or individual rights. SOCO specifies the intuitively appropriate solution in such cases: a *general* solution concept solves the Liberal Paradox. The multi-issue framework also enables us to examine the converse problem of collective choice: getting individual choices to reflect a given collective decision.

Heavily formal, these chapters presuppose some experience with abstract mathematical reasoning. Even so, hardly any specific mathematical knowledge is required—just some naive set theory, whose language is outlined in an appendix—and it is easy to skip proofs. Some of these proofs are long and tedious—paragons of inelegance—partly because they are about objects and structures for which there was no well-developed body of theory from which to draw useful basic results, and partly because my aim often is to avoid certain familiar simplifying assumptions, notably transitivity assumptions. The theorems proved here should make future theorems about the same topics easier to prove. For example, the equivalences proved at some length in chapters 6, 8, and 9 make it possible to choose among alternative formulations of the same idea according to one's deductive goals, switching from one formulation to another without further argument.

PART I

THE FORMAL REPRESENTATION OF CHOICE PROCESSES

CHAPTER ONE

CHOICE FUNCTIONS

1.1. THE CHOICE-FUNCTION CONCEPT

In much of this study I represent choice processes, individual and collective, by so-called choice functions.

Every choice, individual or collective, is *from* a set of *alternatives*, a range of choice options. These are the possible choices that are feasible, or available, on the occasion of choice—those that the individual or individuals in question have the opportunity and ability to choose on that occasion. Let us call them the *feasible alternatives* (on the given occasion), and their set the *feasible set* (on that occasion). Often feasible sets are called *agendas*.

To avoid certain technical problems I cannot solve, I shall mostly limit my discussion to *finite* feasible sets. I assume each such set is nonempty: there always is something to choose. And I assume the feasible alternatives on any occasion are mutually exclusive: whoever is doing the choosing can choose any one of them, but no more than one.

I assume that a process (rule, mechanism, criterion) for making choices, individual or collective, is applicable to any finite, nonempty subset of some *universal set, A*, of *alternatives*. So the family S of *potential feasible sets* may be defined as follows:

DEFINITION S = *the family of finite, nonempty subsets of* A.

Note that the members of A are *alternatives* but that the members of S are *sets* of alternatives—subsets of A.

The choice process itself may be represented by a function C—a *choice function*—fulfilling the following two axioms:

AXIOM C1 $C: S \to S$.

AXIOM C2 $C(\alpha) \subseteq \alpha$ *for all* $\alpha \in S$.

Together these axioms say that C assigns a nonempty subset to every set in S—to every finite, nonempty subset of A.

C represents what I call (for want of a better label) a *choice process*—a rule, procedure, institutional mechanism, or criterion, or a set of tastes, values, goals, or behavioral dispositions, that might govern the choices of some individual or group of individuals from finite subsets of A. For $\alpha \in S$, $C(\alpha)$ comprises those members of α that choosers governed by the given process might choose when α is the feasible set. $C(\alpha)$ is the *choice set* from α. Its members are the *choosable elements* of α.

1.2. SOME QUESTIONS ABOUT THIS APPROACH

Question: You take it for granted that choosers are confronted by feasible sets, whence they must choose. But how are these sets determined? How do the choosers, or those who would explain, predict, evaluate, or prescribe their choices, decide what the feasible alternatives are?

Reply: I fully agree that identifying the feasible set is an important, interesting, neglected part of the total choice process. It is the subject of chapter 10. There, as it happens, the choice-function idea still proves to be a convenient representation of choice processes.

Question: You assume choosers invariably choose *some* feasible alternative, an assumption built into Axiom C1, which says $C(\alpha)$ always belong to S, hence is nonempty. Is it not possible to choose *nothing*—to refrain or abstain from choosing?

Reply: Not as I use the verb *to choose*. My assertion that choosers invariably choose *some* feasible alternative is merely a terminological convention. If it would be correct, in common parlance, to say there are n feasible alternatives and choosers need not choose any of them, I should concede the possibility thus envisaged but describe it differently. In my sense, there are $n + 1$ feasible alternatives and choosers will perforce choose one of them. They are the original n alternatives plus the alternative of choosing none of those n. And instead of saying choosers can sometimes refrain or abstain from choosing anything, I should say choosers can sometimes refrain or abstain from other, more active options.

Question: You assume the alternatives constituting any feasible set are mutually exclusive: any one of them can be chosen, but no more than one. Is this not flatly false in many cases? Consider an election in a four-member constituency with twenty candidates. Must not the electorate choose fully four of the feasible alternatives?

Reply: My assertion that choosers cannot choose more than one alternative, like my assertion that they must choose at least one, is just a terminological convention: that is how I use the word *alternative.* In the electoral example, the feasible alternatives *in my sense* are not the twenty candidates but the 4,845 fourfold combinations of those candidates.

Question: Though not necessarily the feasible set on any occasion, A contains every potential feasible set. Beyond that, just what is A supposed to comprise?

Reply: For present purposes it is enough simply to interpret A as the union over the domain of C—the union of those potential feasible sets to which the process represented by C is applicable.

It may help to think of A as the set of all conceivable or imaginable or logically possible choice alternatives, not just the feasible ones on this or that occasion. Somewhat more precisely: A choice process normally is applicable only to alternatives of some broad but specifiable type—possible legislative enactments, allocations of certain goods, people who fulfill certain qualifications, or whatever; A may be regarded as the type in question. More precisely still: Being mutually exclusive, the alternatives to which a choice process is applicable can be regarded as complete descriptions of the world in certain specifiable terms, or as vector representations of possible states of the world that differ in certain specifiable dimensions; A may then be regarded as the set of all such world descriptions or state vectors. This sketch of an interpretation makes it likely that A is infinite, although each potential feasible set—each member of S—is finite.

Question: Why is $C(\alpha)$ defined for *all* finite, nonempty subsets of A? Why do you assume the process represented by C is applicable to *every* finite range of alternatives of some sort?

Reply: I make this assumption for mathematical convenience.

Admittedly it limits the scope of my discussion somewhat. But it is widely satisfied all the same. Although restricted in the *kinds of alternatives* to which they are applicable, extant collective-choice processes often are applicable to *any finite range* of alternatives of the appropriate kind.

There are exceptions, to be sure. Some collective-choice processes, notably Simple Majority Rule and the so-called special-majority rules (Two-Thirds, Three-Quarters, and the like), are applicable only to *two-element* feasible sets. But these processes normally are parts of more complex, *sequential* choice processes. And although each stage in such a process involves a two-element feasible set, the process as a whole is applicable to feasible sets of arbitrary finite cardinality; whereof see chapter 9. Some choice processes apparently apply only to feasible sets containing a *status quo* alternative. But as before, such a set normally can be regarded as the feasible set at some stage in a (possibly long and gappy) sequential choice process, the status quo at each stage being the alternative chosen at the previous stage. And once again, the serial process *as a whole* is applicable to any set whose members are all of an appropriate kind.

Question: Because the members of any potential feasible set α are mutually exclusive, two or more of them cannot simultaneously be chosen from α. Yet Axioms C1 and C2 allow $C(\alpha)$ to have more than one member. Why? Why not postulate that $C(\alpha)$ is a singleton—or, more simply, a member of α rather than a subset of α?

Reply: When $C(\alpha)$ has more than one member, the process represented by C has narrowed the choice from α to $C(\alpha)$ but has not narrowed it further, leaving the final choice from $C(\alpha)$ to other factors, processes, or acts, sometimes naturally described as chance, random processes, or arbitrary choices.

If we regard the process represented by C as a rule, principle, or criterion that prohibits some choices, permits others, and prescribes some of those it permits, we can say that $C(\alpha)$ comprises the *permitted* choices from α: given that α is the feasible set, the process in question prohibits the choice of anything in $\alpha - C(\alpha)$, permits the choice of anything in $C(\alpha)$, and, if $C(\alpha)$ is a singleton, prescribes the choice of the sole alternative therein. Multimember choice sets, particularly large ones, are the mark of the relatively *permissive* choice processes. Single-member or small choice sets mark a relatively *restrictive* choice process.

If we regard the process represented by C as a mechanism or set of behavioral dispositions that causes or produces certain results, we can say, when $C(\alpha)$ has more than one member, that the process *partly determines* the choice from α, causing it to fall within a certain range of alternatives (those constituting $C(\alpha)$) without causing it to be any particular member of that range. Multimember choice sets, especially large ones, are the mark of a relatively *indeterministic* choice process. Single-member or small choice sets mark a relatively *deterministic* choice process—totally deterministic, if $C(\alpha)$ is always a singleton.

Although itself somewhat permissive or indeterministic, the process represented by C could still be *part* of a fully restrictive or deterministic choice process, the part we know about or find interesting or important or otherwise have reason to study separately at the moment. If it is the part we know about, then a multimember $C(\alpha)$ reflects incomplete predictive power: we can predict that a choice from α will be one of the several alternatives in $C(\alpha)$ but cannot predict which one, regarding the final choice from $C(\alpha)$ as random, as a matter of chance. If it is the part we find interesting or important, then once the choice from α has been narrowed to $C(\alpha)$, the further choice from $C(\alpha)$ will strike us as somehow arbitrary, meriting little, if any, explanation or justification.

Question: To represent a choice process by a function like C is to represent it as a function solely of feasible sets. Do collective choices not depend on other factors—on the way people vote, for example?

Reply: Certainly. But when I represent a collective-choice process by a choice function, I assume that all choice-determining factors other than the feasible set are held constant (remain fixed) throughout the context of my discussion. In the sequel, I sometimes treat factors other than the feasible set—individuals' perferences, for example—as variables upon which collective choices depend (or may depend). The choice process under consideration may then be represented by a functional $\mathbb{C}$ (a function whose values are themselves functions) of the following sort: $\mathbb{C}$ is defined on a set K of vectors whose dimensions represent the additional choice-determining factors, and the value, $\mathbb{C}^v$, of $\mathbb{C}$ at each vector v in K is a choice function on S—a function satisfying Axioms C1 and C2. In short:

$$\mathbb{C} : K \rightarrow \{f | f : S \rightarrow S \text{ and } f(\alpha) \subseteq \alpha \text{ for all } \alpha \in S\}.$$

More intuitively, consider a vector v in K: it is a possible specification of all choice-determining factors other than the feasible set—voters' preferences, parliamentary categories of motions, entrail configurations, or whatnot. And consider a set α in S—a potential feasible set. The result, $\mathbb{C}^v$, of applying the function $\mathbb{C}$ to v (the value of $\mathbb{C}$ at v) is itself a function—a choice function—which, when applied to α, yields the choice set $\mathbb{C}^v(\alpha)$—the set of permissible choices in the situation characterized by v and α. My C is just $\mathbb{C}^v$ for an arbitrary but fixed value of v.

To represent a choice process by a function of soandsos but not of suchandsuches is to say nothing about the nature of the process; it is not to imply that the process actually depends on soandsos but not suchandsuches. Possibly the process really depends on suchandsuches but the suchandsuch factor is held fixed—not allowed to vary—for immediate purposes of discussion; then the representing function is a *covert function* of suchandsuches. Possibly the process really does not depend on soandsos, in which case the representing function is a *constant* function of soandsos—one whose value remains invariant under all possible variations of the soandso variable. (It is not necessarily gratuitous to use such a constant function because the fact that it *is* a constant function of its soandso parameter could well be an unexpected *conclusion*, not an otiose *premise*.)

ANYTHING CAN BE REPRESENTED AS A FUNCTION OF ANYTHING ELSE. When a mathematical investigator represents *X*'s as a function of *Y*'s but not of *Z*'s, he tells you something about his interests, not about *X*'s, *Y*'s, or *Z*'s. When I represent certain collective choices as a function of soandsos but not of suchandsuches, all I tell you thereby is that I am currently investigating the relationship between those choices and soandsos but not suchandsuches. I do not tell you that those choices really depend on soandsos but not on suchandsuches.

1.3. CLASSICAL CONDITIONS OF "RATIONAL" CHOICE

Often it is said there is something anomalous about a choice function, or the process represented thereby, that fails to fulfill certain conditions of "rationality," or "rational choice." These conditions link choice functions to underlying "preference" and "indifference" relations, imputing a simple, familiar, intuitively attractive structure to such functions and relations. Sometimes they are proferred as definitional requirements that any choice function, properly so-called, must meet.

It is generally and pretty much uncritically assumed that these conditions hold for individual-choice processes (or, if you prefer, for the choice functions that represent such processes). Whether they hold and whether they should hold for certain or all *collective*-choice processes are matters of considerable controversy, occupying many of the pages to follow. Whether they hold for a given collective-choice process unquestionably makes a significant difference to the nature of that process and to some of the choices made thereby, as you will see in Part II. And whether individual participation in collective-choice processes is characterizable in terms of rational individual choice, classically construed, also makes a difference.

Let P, the "preference" relation corresponding to C, be a binary relation on A. Define *indifference, I,* as the absence of preference in either direction between members of A:

xIy iff $x, y \in A$ but neither xPy nor yPx.

I shall say more about the nature of preference and indifference after examining the roles these relations play in the traditional assumptions of rationality.

The most fundamental of these assumptions is that $C(\alpha)$ is identically the set of *P-undominated elements* of α—those elements to which no other elements bear P:

BINARY-CHOICE PROPERTY (BICH) $C(\alpha) \equiv \{x \in \alpha \mid yPx \textit{ for no } y \in \alpha\}$.

So a choice function satisfies BICH if it always picks the most preferred of the feasible alternatives. This is the bare bones of the familiar idea that to choose rationally is to *maximize* something.

Since, for every $\alpha \in S$, $C(\alpha)$ has at least one member (by Axiom C1), BICH implies that every $\alpha \in S$ has at least one P-undominated member. And this is equivalent to:

P-ACYCLICITY *Not* $x_1Px_2P \cdots x_{k-1}Px_kPx_1$,

that is:

There is no *P*-cycle,

where:

DEFINITION α *is a* ρ*-cycle iff* ρ *is a binary relation, and for some integer k and objects* $x_1, \ldots, x_k$, $\alpha = \{x_1, \ldots, x_k\}$ *and* $x_1 \rho x_2 \cdots x_{k-1} \rho x_k \rho x_1$.

(Note that I have defined cycles for *arbitrary* binary relations ρ, not just for the particular relation *P*.) *Proof* that P-Acyclicity is equivalent to the condition that every $\alpha \in S$ has at least one *P*-undominated member: Were there a *P*-cycle, it would lack a *P*-undominated member. Conversely, were there an $\alpha \in S$ without a *P*-undominated member, we could construct a *P*-cycle this way: Enumerate the elements of α, and let x_1 be the first element in the enumeration, x_2 the first element that bears *P* to x_1, x_3 the first element that bears *P* to x_2, and so on. Since α lacks a *P*-undominated member, the sequence $(x_1, x_2, x_3, \ldots)$ is infinite, and thus, since α is finite, the sequence must have repetitions: for some $i, j, j > i$ although $x_j = x_i$. Hence, $x_j P x_{j-1} \cdots x_{i+1} P x_j$, contrary to P-Acyclicity.

BICH plus Axiom C1 and the definition of *I* also imply:

P-ASYMMETRY *Not both xPy and yPx.*

I-REFLEXIVITY *xIx for all* $x \in A$.

I-SYMMETRY *If xIy then yIx.*

PI-TRICHOTOMY *If* $x, y \in A$, *exactly one of the following: xPy, yPx, or xIy.*

We get a stronger rationality condition if we conjoin BICH to:

P-TRANSITIVITY *If xPyPz then xPz,*

and a still stronger one if we add:

I-TRANSITIVITY *If xIyIz then xIz.*

Assuming P-Asymmetry, *P* is a *suborder* if it satisfies P-Acyclicity, a *strict partial order* (a partial order in the asymmetric sense) if it satisfies P-Transitivity, and a *strict weak ordering of A* (a weak ordering of *A* in the asymmetric sense) if it satisfies both P-Transitivity and I-Transitivity (with I defined as above). The conjunction of P-Transitivity and I-Transitivity—*P + I-Transitivity*, for short—is equivalent to:

P ∪ I-TRANSITIVITY *If xP ∪ Iy and yP ∪ Iz then xP ∪ Iz,*

and to:

NEGATIVE P-TRANSITIVITY *For all x, y, z ∈ A, if not xPy and not yPz then not xPz.*

Like BICH, the combination of P-Asymmetry and P-Transitivity is stronger than P-Acyclicity: it implies but does not follow from P-Acyclicity.

There are innumerable transitivity conditions involving P and I that are stronger than P-Transitivity yet weaker than P + I-Transitivity. The following two are especially interesting:

PIP-TRANSITIVITY *If xPyIzPw then xPw (P is an* interval order *of A).*

PIP + IPP-TRANSITIVITY *If xPyIzPw then xPw, and if xIyPzPw then xPw (P is a* semiorder *of A).*

These conditions are designed to allow such violations of P + I-Transitivity as could be due to limited powers of discrimination—to nonnoticeable differences adding up to noticeable ones. For example, we could have *xIyIzPx* (contrary to P + I-Transitivity but consistently with PIP + IPP-Transitivity) if a nonnoticeable difference between x and y and a nonnoticeable difference between y and z added up to a noticeable difference between x and z. Assuming A is countable, P + I-Transitivity is equivalent to:

For some real-valued function u on A and for all $x, y \in A$, xPy iff $u(x) > u(y)$,

which says P can be interpreted as the relation of one element of A to another when the first has a greater "utility" than the second; while PIP-Transitivity is equivalent to:

For some real-valued functions u and d on A, with d (the "discriminability" measure) positive, and for all $x, y \in A$, xPy iff $u(x) > u(y) + d(y)$, which says P can be interpreted as the relation of one element of A to another when the first has a *noticeably* greater utility than the second; and PIP + IPP-Transitivity is equivalent to:

For some real-valued function u on A and for all $x, y \in A$, xPy iff $u(x) > u(y) + 1$,

which says P can be so interpreted that (1) P is the relation of one element of A to another when the first has a noticeably greater utility than the second and (2) a utility difference noticeable at any level of satisfaction is noticeable at all levels.

Some discussions of individual and collective choice assume that P is not merely a weak ordering of A but a *linear* ordering. This means P satisfies not only P + I-Transitivity but also:

P-CONNEXITY *For all $x, y \in A$, if $x \neq y$ then either xPy or yPx (P is a* tournament *on A).*

In effect, this proscribes all indifferences between two alternatives. Although P-Connexity alone does not imply P-Transitivity, it does so in conjunction with P-Acyclicity, and it does by itself imply I-Transitivity, given my definition of I. So in the presence P-Connexity, the relatively strong P + I-Transitivity is equivalent to its weak cousin, P-Acyclicity.

1.4. WHAT IS THIS THING CALLED PREFERENCE?

Just what are P and I? Not necessarily preference and indifference in any colloquial sense of these terms. Theorists who postulate BICH and the other conditions in models of individual or collective choice often would contend—and more often would *need* to contend—no more than this: The choosers in question choose as though they were *maximizing* something—*something or other*, not necessarily anything in particular, not necessarily preference satisfaction in any antecedently understood sense of preference.

I have defined indifference (I) as the absence of preference (P) in either direction between members of A. One might object to this (or to PI-Trichotomy) on the ground that two alternatives could be *incomparable*, hence not indifferent yet not linked by P.

I find it convenient, however, to *interpret* indifference so that any two incomparable alternatives (if such there be and whatever that be) automatically qualify as indifferent. This terminological fiat has its price: it can produce violations of I-Transitivity. For if x and y are incomparable and y and z are incomparable, that is no guarantee x and z are incomparable; z might be preferred to x, in which case $xIyIzPx$, contrary to I-Transitivity.

In a way, the meaning of preference (P) is implicit in BICH. Given Axioms C1 and C2, and given that P is a binary relation on A, BICH implies that xPy holds if and only if x and y are distinct elements of A and $C(\{x, y\}) = \{x\}$. This means P is the relation of one alternative to another when C chooses the first and it alone from their pair set—when choosers governed by C would choose the first over the second, given that they alone were feasible.

Proof. Assume Axioms C1 and C2 and BICH, and assume that $P \subseteq A^2$. Suppose xPy. Then $x, y \in A$; and by BICH, $y \notin C(\{x, y\})$. So by Axioms C1 and C2, $C(\{x, y\}) = \{x\}$, and thus $x \neq y$. Conversely, suppose $x, y \in A$, $x \neq y$, and $C(\{x, y\}) = \{x\}$; to deduce that xPy. By hypothesis, $y \notin C(\{x, y\})$. So by BICH, either xPy or yPy. But by BICH again, if yPy then $C(\{y\}) = \emptyset$ (the empty set), contrary to Axiom C1. Hence, xPy. ■

Notice five things about this consequence of BICH:

Thing 1. In effect, it *defines* P in terms of C—just as BICH itself, in effect, defines C in terms of P.

Thing 2. So defined, P represents the restriction of C to pair sets: given P, we can always reckon $C(\{x, y\})$; it is $\{x\}$ or $\{y\}$ or $\{x, y\}$ according as xPy or yPx or neither. Hence:

Thing 3. BICH reduces the problem of choosing from arbitrary finite sets of alternatives to that of choosing from pair sets.

Thing 4. Depending on the sort of choice process represented by C, P may or may not be the relation of preference, *ordinarily so-called*, of a subject whose choices are governed by C. In common parlance, it makes sense to say someone would choose one alternative over another, given a choice between the two, but does not prefer the former to the latter. Example: I choose to pay my taxes, although I prefer not to pay. I pay because it is my *duty*, not because I *prefer* paying to not paying. Preference, ordinarily so called, is just one factor (or cluster of factors) motivating or explaining choices. Hence:

Thing 5. In no sense does BICH explain choices in terms of *other* things, called preferences. At most it "explains" some choices in terms of other *choices*.

Henceforth, let us define P in the above manner, *whether or not BICH is assumed.*

DEFINITION *xPy iff* $x, y \in A$, $x \neq y$, *and* $C(\{x, y\}) = \{x\}$.

In short, *preference* is *pairwise choice.*

Because BICH implies this definition (given that P is a binary relation on A), the relation P, so defined, is the one and only relation on A that could satisfy BICH: if BICH holds for any relation on A, that relation must be this one. So these two formulations of BICH (the first of which was given earlier) are equivalent:

For every $\alpha \in S$, $C(\alpha) = \{x \in \alpha \mid yPx \text{ for no } y \in \alpha\}$.

There exists a binary relation ρ on A such that, for every $\alpha \in S$, $C(\alpha) = \{x \in \alpha \mid y\rho x \text{ for no } y \in a\}$.

Assuming Axioms C1 and C2 and the definition of indifference as the absence of strict preference in either direction between members of A, the definition of P implies that xIy holds if and only if $x, y \in A$ and $C(\{x, y\}) = \{x, y\}$. This means I is the relation of one alternative to itself or another when C allows either to be chosen—when, according to C, it is a *matter of indifference* which is chosen—given that they alone are feasible. Let us henceforth eschew the earlier definition of I and adopt this one:

DEFINITION *xIy iff* $C(\{x, y\}) = \{x, y\}$.

By themselves, the definitions of P and I have four intuitively expected consequences:

P-ASYMMETRY *If xPy, not yPx.*

P-IRREFLEXIVITY *not xPx.*

I-SYMMETRY *If xIy, yIx.*

PI-DISJOINTNESS *Not both xPy and xIy.*

Combined with Axioms C1 and C2 and the definition of *S*, the definitions of *P* and *I* further imply:

I-REFLEXIVITY *xIx for all x ∈ A.*

PI-TRICHOTOMY *For all x, y ∈ A, exactly one of the following: xPy, yPx, or xIy.*

STRONG P ∪ I-CONNEXITY *For all x, y ∈ A, either xP ∪ Iy or yP ∪ Ix.*

When investigating properties of *C, P,* and *I*, I shall always assume Axioms C1 and C2, the definitions of *S, P,* and *I*, and therewith the consequences just listed.

CHAPTER TWO

INDIVIDUAL PARTICIPATION

2.1. COLLECTIVE CHOICE AS A FUNCTION OF INDIVIDUAL PREFERENCE

Collective-choice processes often are represented as functions of individual preferences—the preferences of the participants in the process.

Let there be *n individual participants*, Messrs. 1, 2, . . . , *n*. They may be corporate entities rather than human persons. I call them individuals because I ignore their internal composition.

Let N be the *set of individuals*—or, strictly speaking, of their indices:

DEFINITION $N = \{1, 2, \ldots, n\}$.

Individual Preferences

Each individual has a preference relation on A. Sometimes I presuppose a fixed set of individual preference relations, $P_1, \ldots, P_n$, satisfying:

AXIOM P_i *$P_1, \ldots, P_n$ are asymmetric binary relations on A.*

More often I allow individual preferences to vary, in which case *PROF* is a set of *preference profiles*, each an ordered n-tuple of relations $\rho_1, \ldots, \rho_n$, representing a possible situation in which ρ_1 is Mr. 1's preference relation, ρ_2 Mr. 2's, and so on. Any two preference profiles represent situations that differ only in these relations and perhaps other factors that depend on these relations. *PROF* represents the range of possible situations to which the collective-choice process under consideration is applicable. Note that a profile (a member of *PROF*) is a *combination* of preference relations, one for each individual, not a *single* preference relation.

The following three definitions provide useful language for talking about preference profiles:

DEFINITION $xP^v_r y$ *iff, for some* $\rho_1, \ldots, \rho_n$, $v = (\rho_1, \ldots, \rho_n) \in PROF$ *and either* (1) $r \in N$ *and* $x\rho_r y$, *or else* (2) $r \subseteq N$ *and* $x\rho_i y$ *for every* $i \in r$.

So $xP^v_r y$ holds if, in the v-situation, r is either an individual who prefers alternative x to alternative y or a group of individuals whose members all prefer x to y.

DEFINITION *If* α *is a set and* ρ *a binary relation then*

$$\rho/\alpha = \rho \cap \alpha^2.$$

If α *is a set and* ρ *a vector of binary relations* $\rho_1, \ldots, \rho_k$ *then*

$$\rho/\alpha = (\rho_1/\alpha, \rho_2/\alpha, \ldots, \rho_k/\alpha).$$

Otherwise, $\rho/\alpha = \emptyset$.

If α is a set and ρ a binary relation or vector of binary relations, then ρ/α is the *restriction* of ρ to α (in one prevalent sense of that term). So if ρ is the parent-offspring relation and α is the set of females then ρ/α is the mother-daughter relation. If Messrs. 1, 2, . . . , k are the judges in a tasting of 1983 California chardonnays, ρ_i the relation of gustatory superiority in Mr. i's judgment, and α the set of chardonnays in the tasting, then $(\rho_1, \ldots, \rho_k)/\alpha$ comprises the judges' rankings of these particular wines.

DEFINITION (a) α *is* completely free *iff* $\alpha \subseteq A$ *and, for all* (*strict*) weak *orderings* $\rho_1, \ldots, \rho_n$ *of* α, *there is a* $v \in PROF$ *for which* $v/\alpha = (\rho_1, \ldots, \rho_n)$. (b) α *is* linearly free *iff* $\alpha \subseteq A$ *and, for all* linear *orderings* $\rho_1, \ldots, \rho_n$ *of* α, *there is a* $v \in PROF$ *for which* $v/\alpha = (\rho_1, \ldots, \rho_n)$.

So if α is a *completely* free set, every possible combination of *weak* orderings of α is embedded in some profile belonging to *PROF*. And if

α is a *linearly* free set, every possible combination of *linear* orderings of α is embedded in some profile belonging to *PROF*. In either case, unless $\alpha = A$, the profile in question can differ from the given combination of orderings of α in that it represents individual preferences on the whole set A, not just on α. Because every linear ordering of a set is also a weak ordering, but not conversely, complete freedom is a stronger condition on a set than linear freedom.

I shall always assume:

> AXIOM *PROF* *Every member of PROF is an ordered n-tuple of asymmetric binary relations on A.*

From time to time I shall examine further assumptions about the membership of *PROF*.

One such assumption that is frequently made is the following:

> UNRESTRICTED DOMAIN (a) *Every vector in PROF comprises only weak orderings of A.* (b) *A is completely free, that is, every ordered n-tuple of weak orderings of A belongs to PROF.*

In studies of collective choice, (a) is almost always assumed but hardly ever used. I shall not assume it.

In place of (b), I shall often assume for various values of k:

> k-DOMAIN $|A| \geq k$, *and every k-element subset of A is linearly free.*

If A is large compared with k—especially if A is infinite and k finite—k-Domain is significantly less demanding than (b) in this respect: Where (b) requires *PROF* to include profiles in which Messrs. 1, 2, . . . , n have as many levels of discrimination as there are alternatives in A, k-Domain just requires *PROF* to include profiles in which Messrs 1, 2, . . . , n have k levels of discrimination. Thus, depending on the sizes of k and A and on the nature of the alternatives in A, (b) could well be psychologically unrealistic compared with k-Domain.

Even if $|A| = k$, k-Domain is significantly less demanding than (b) in another respect: Unlike (b), k-Domain does not require *PROF* to contain profiles of *nonlinear* individual preferences. This is important because virtually all real-world voting procedures in which voters rank alternatives on a ballot instead of merely specifying their first choices require linear rankings. (Alternatives not expressly ranked on a ballot are usually treated as though they were mutually indifferent and dispreferred to those expressly ranked. But arbitrary nonlinear weak orderings are disallowed.) Unlike some studies of collective choice, however, mine does *not assume* linearity; k-Domain is neutral on the issue.

Like (b), the many instances of k-Domain (differing in the value of k) ensure a measure of preferential variety among and within the profiles comprised in *PROF*. Because these domain conditions are used in chapter 3 to obtain some surprisingly strong results, it is well to control their strength as much as possible by assuming linear rather than complete freedom, of small rather than large sets.

Collective Choice, Preference, and Indifference

If we wish to represent a collective-choice process by a function of individual-preference relations, we can use a functional $\mathbb{C}$ that assigns to each $v \in PROF$ a corresponding choice function $\mathbb{C}^v$—a function $\mathbb{C}^v$ satisfying Axioms C1 and C2. The corresponding pairwise (two-alternative) choice process may then be represented by a function $\mathbb{P}$ that assigns to each $v \in PROF$ the preference relation corresponding to $\mathbb{C}^v$—the relation $\mathbb{P}^v$ defined as follows:

$$x \, \mathbb{P}^v y \text{ iff } x, y \in A, x \neq y, \text{ and } \mathbb{C}^v(\{x, y\}) = \{x\}.$$

Because my formal results are cast in terms of $\mathbb{P}$ rather than $\mathbb{C}$, I shall take $\mathbb{P}$ (along with A and n) as primitive and assume:

AXIOM P $\mathbb{P}$ *is a function on PROF whose value*, $\mathbb{P}^v$, *at each* $v \in PROF$, *is an asymmetric binary relation on A.*

$\mathbb{P}$ is the *collective-preference function*. And for $v \in PROF$, $\mathbb{P}^v$ is the *collective-preference relation* in the v situation.

Corresponding to $\mathbb{P}$ is a *collective-indifference* function, defined thus:

DEFINITION $\mathbb{I}$ *is the function on PROF whose value*, $\mathbb{I}^v$, *at each* $v \in PROF$ *is the relation defined by:*

$$x \, \mathbb{I}^v y \text{ iff } x, y \in A \text{ and neither } x\mathbb{P}^v y \text{ nor } y\mathbb{P}^v x.$$

Decisive Sets

The following three definitions will help us discuss the power structure embodied in $\mathbb{P}$:

DEFINITION *g is* decisive *for x vs. y iff* $g \subseteq N$, $x, y \in A$, $x \neq y$, *and for every* $v \in PROF$, *if* $xP^v_g y$ *but* $yP^v_{N-g}x$ *then* $x\mathbb{P}^v y$.

DEFINITION *g is* nearly decisive *for x vs. y iff* $g \subseteq N$, $x, y \in A$, $x \neq y$, *and for every* $v \in PROF$, *if* $xP^v_g y$ *but* $yP^v_{N-g}y$ *then not* $y\mathbb{P}^v x$.

So a group *g* is *decisive* for an alternative *x* vs. another alternative *y* if *x* is collectively preferred to *y* under every profile in which the *g*'s all prefer *x* to *y* and the non-*g*'s all prefer *y* to *x*. And *g* is *nearly* decisive for *x* vs. *y* if *y* is not collectively preferred to *x*—if, in other words, *x* is collectively preferred or indifferent to *y*—under every such profile. Decisiveness implies near decisiveness but not conversely: a group that is *nearly* decisive for *x* vs. *y* can always dictate a collective *preference or indifference* for *x* over *y* (blocking a collective preference for *y* over *x*) in the face of unanimous opposition by nonmembers, but only a group that is *decisive* for *x* vs. *y* can always dictate a (strict) collective *preference* for *x* over *y* in such a case.

DEFINITION *g is* superdecisive *iff* $g \subseteq N$ *and the following two conditions hold for all distinct* $x, y \in A$ *and all* $v \in PROF$:

(1) *Every h for which* $g \subseteq h \subseteq N$ *is decisive for x vs. y.*

(2) *If every three-element subset of A is completely free and* $xP^v_g y$ *then* $x\mathbb{P}^v y$.

A superdecisive group can dictate the collective preference, not only for some specified pair of distinct alternatives, but for every pair, and not only when unanimously opposed by nonmembers, but also when some nonmembers share the group's preference; and this power extends to situations in which some individuals have nonlinear preferences, so long as every set of three alternatives is completely free.

2.2. PREFERENCE INTENSITIES

A preference profile tells us who prefers what to what. It tells us nothing about the *intensity* of those preferences. It could not tell us, for example, that Mr. 3 prefers the Vegetarian to the Flat Earth candidate 4.87 times as intensely as Mr. 18 prefers the Flat Earther to the Vegetarian. Some theorists hold, though, that collective choices ought to reflect such preference-intensity information when it is available.

The preference-intensity information most important for collective choice consists of *numerically specific interpersonal comparisons* of preference intensity, such as the comparison above involving Mr. 3 and Mr. 18. These contrast with numerically *in*specific interpersonal comparisons, like "Mr. 3 prefers the Vegetarian to the Flat Earther more intensely than Mr. 18 prefers the Flat Earther to the Vegetarian" (precisely *how much* more?), and with in*tra*personal comparisons, such as "Mr. 3 prefers the Vegetarian to the Flat Earther more intensely than [32.6 times as intensely as] *he* prefers the Flat Earther to the Prohibitionist." All these are *difference* comparisons, to be distinguished from *level* comparisons, like "Mr. 3 favors the Vegetarian more intensely than [8.387 times as intensely as] Mr. 18 favors the Flat Earther," which are needed if we wish to take account of the way preference satisfaction is *distributed*.

When discussing preference intensities, it is customary to represent individual preferences by real-valued functions on A rather than binary relations on A—by "utility" measures rather than preference relations. An ordered n-tuple $(u_1, \ldots, u_n)$ of such functions—a *utility profile*—represents a situation in which, for example, if $u_i(u) = 1$, $u_i(y) = 0$, $u_j(x) = 14$, and $u_j(y) = 17$, then Mr. j prefers y to x three times as intensely as Mr. i prefers x to y.

To represent collective choice as a function of preference profiles rather than utility profiles is not to imply that collective choices depend on no preference-intensity information. Remember: Anything can be represented as a function of anything else. Although an explicit function only of individual-preference relations, $\mathbb{P}$ might be a covert function of interpersonally significant individual utility functions. Specifically, $\mathbb{P}^v$ can be interpreted, for each $v \in PROF$, as the relation of one alternative to another when, in the v situation, those who prefer the former to the latter have a greater aggregate preference intensity than those who prefer the latter to the former.

To prove the point to skeptics, let f be a function that assigns to each preference profile v a utility profile $f(v) = (u_1, \ldots, u_n)$ that is *compatible* with v in this sense: for all $x, y \in A$ and $i \in N$, xP^v_iy iff $u_i(x) > u_i(y)$; interpret $f(v)$ as the utility profile that correctly represents people's preference intensities in the v situation. Now define $\mathbb{P}$ on *PROF* this way:

$x\mathbb{P}^vy$ iff $x, y \in A$, and for some $u_1, \ldots, u_n, f(v) = (u_1, \ldots, u_n)$

and $\sum_{i=1}^{n} u_i(x) > \sum_{i=1}^{n} u_i(y)$.

Then for every $v \in$ *PROF*, $x\mathrm{P}^v y$ holds if, and only if, $x, y \in A$ and those who prefer x to y in the v situation have a greater aggregate preference intensity than those who prefer y to x in that situation.

In the sequel, I shall give very little attention to collective-choice processes based on preference intensities, for two reasons:

First, such processes are thoroughly unrealistic. None is used in the real world. We do not even know how, in principle, to measure preference intensity. And if we did know how, doing so would likely be impracticable: if the relevant information were not obtained by polling people about their feelings (or a similar procedure), the cost of obtaining it (the moral as well as the economic cost) probably would be prohibitive, and if it were indeed obtained by such polling, most people would respond so as to give evidence of more intense preferences than they really had.

Second, it is questionable whether collective choices *should* be based on preference-intensity information, even when such information is ready to hand. Partisans of the contrary view apparently feel that the purpose (proper purpose, most likely purpose) of instituting a collective-choice process is to satisfy participants' preferences, on the whole, to the greatest degree possible. And this evidently requires that collective choices reflect preference intensities whenever possible.

But in the first place, there are other purposes served by instituting collective-choice processes—worthier and more likely purposes, I contend—and these do not obviously require that collective choices reflect preference intensities, even when preference-intensity data are available. One such purpose is to distribute power widely, thereby preventing concentrations of power and so minimizing the abuse of power. Another purpose is to broaden the pool of ideas by which choices are informed. Yet another is to enhance people's sense of participation in, and therewith their allegiance to, the institutions that govern them. Perhaps the most important purpose is to institutionalize such power shifting, or governmental change, as would otherwise occur in a more violent, less predictable manner. For none of these purposes would interpersonally comparable preference-intensity information be especially useful.

Nor is it fair, generally, to base collective choices on preference intensities, even when these are known. To do so is to attach undue weight to the preferences of those who are mercurial, greedy, bigoted, lustful, earnest, meddlesome, and the like, thereby bringing about inequitable distributions of benefits. Suppose Crusoe and Friday have contributed equally to the day's catch of fish and have equal appetites and similar metabolisms but Crusoe is greedier: unlike Friday, he gets a kick out of the mere perception that he has a larger share of fish. The most reasonable allocation of fish surely would be an equal division of the day's catch. But an allocation based on preference intensities (one designed to maximize aggregate preference satisfaction, possibly discounted by some index of dispersion) would give the lion's share to Crusoe, unfairly rewarding his greed.

That Crusoe's satisfaction depended on a perceived comparison of his share with Friday's was inessential. Just suppose Crusoe is not greedy but gluttonous: he always wants to eat more at a sitting than can be justified on grounds of nutrition, gustatory appreciation, or post-prandial comfort, and he always becomes very unhappy if he does not get all he wants. As before, an allocation of fish based on preference satisfaction would unfairly give Crusoe more than Friday. Or suppose Crusoe and Friday are equally strong and skilled fishermen but Crusoe is lazier. Then an allocation of fishing chores based on preference satisfaction would unfairly assign less work to Crusoe than to Friday—assuming Friday is not especially resentful.

Those who would base governmental choices on preference intensities feel, I suspect, that such choices should maximize something plausibly called *social welfare*. But even if welfare maximization were the proper business of government (a fashionable but questionable view), that goal would be very different from satisfying people's *preferences*, as the Crusoe-Friday example shows.

In fact, satisfying a *single individual's* preferences is very different from promoting *his personal* welfare. Say my daughter and I both need medical treatment badly and I can afford the treatment for either one of us but not both. Being a good father, I prefer that my daughter get the treatment she needs, hence that I not get the treatment I need. But from this it does not follow that *her* getting the treatment *she* needs—or *my not* getting the treatment *I* need—would promote *my* welfare. Satisfying my preferences and promoting my welfare are very different when, as in this example, my preferences are altruistic rather than self-regarding. To cite an example of a very different sort, my preference for

1961 Château La Tour over Vinegro Vinyards Dago Red gives you no license to infer that the former affects my welfare more favorably than the latter. You may not infer that the La Tour is *better for* me, only that *I like* it better.

2.3. PREFERENCES INVOLVING INFEASIBLE ALTERNATIVES

As voting is commonly practiced, voters express preferences for *feasible* alternatives only: voting rules are independent of individual preferences regarding infeasible alternatives. As a condition on $\mathbb{C}$, this means:

> INDEPENDENCE OF IRRELEVANT ALTERNATIVES *If* $\alpha \in S$, v, $w \in PROF$, *and* $v/\alpha = w/\alpha$, *then* $\mathbb{C}^v(\alpha) = \mathbb{C}^w(\alpha)$.

When restricted to the two-alternative case, this condition can be stated in terms of $\mathbb{P}$ rather than $\mathbb{C}$:

> BINARY INDEPENDENCE *If* v, $w \in PROF$, $v/\{x, y\} = w/\{x, y\}$, *and* $x\mathbb{P}^v y$, *then* $x\mathbb{P}^w y$.

Kenneth Arrow first formulated and named the Independence of Irrelevant Alternatives in his classic 1952 study of collective choice. The condition says that if individual preferences among the feasible alternatives (the members of α) are the same in two situations (if $v/\alpha = w/\alpha$), so are the collective choices ($\mathbb{C}^v(\alpha) = \mathbb{C}^w(\alpha)$), even if individual preferences involving infeasible alternatives (members of $A - \alpha$) differ between the two situations. Binary Independence says this just for two-element feasible sets, which means that if individual preferences between two alternatives (x and y) are the same in one situation as in another (if $v/\{x, y\} = w/\{x, y\}$), so is the collective preference ($x\mathbb{P}^w y$ if $x\mathbb{P}^v y$).

These conditions do not require collective choices to be independent of *all* information other than individual preferences among feasible alternatives, but only of all *preferential* information other than individual preferences among feasible alternatives: $\mathbb{C}$ or $\mathbb{P}$ could be a covert function of nonpreferential information of some sort.

It is Binary Independence rather than the full Independence of Irrelevant Alternatives condition that will play a role in my investigations. It plays its principal role in this theorem:

THEOREM 2.3.1. *Assume Binary Independence. Suppose* $g \in G$, $xP^v_g y$, *and* $yP^v_{N-g}x$. *Then* (a) *if* $x\mathrm{P}^v y$, *g is decisive for x vs. y, and* (b) *if not* $y\mathrm{P}^v x$, *g is nearly decisive for x vs. y.*

This says that if a group of individuals succeeds in dictating (or blocking) a collective preference in *one* situation despite the unanimous opposition of nonmembers, then it has the power to dictate (or block) the same collective preference in *every* situation in which it is unanimously opposed by nonmembers.

Proof. (a) Suppose $xP^w_g y$ and $yP^w_{N-g}x$; to deduce that $x\mathrm{P}^w y$. But $v/\{x, y\} = w/\{x, y\}$. By Binary Independence, then, since $x\mathrm{P}^v y$, $x\mathrm{P}^w y$.

(b) Suppose $xP^w_g y$ and $yP^w_{N-g}x$; to deduce that not $y\mathrm{P}^w x$. But $w/\{x, y\} = v/\{x, y\}$. By Binary Independence, then, $y\mathrm{P}^v x$ if $y\mathrm{P}^w x$. But not $y\mathrm{P}^v x$. So not $y\mathrm{P}^w x$. ■

Independence of Irrelevant Alternatives and therewith Binary Independence are eminently reasonable assumptions to make in a realistic study of collective choice. I know of no real-world collective-choice process that violates either condition. Both formalize the idea that collective choices depend only on such preferential data as could be revealed by *voting.*

Beginning with Arrow, some theorists have argued otherwise, contending that certain familiar voting procedures violate Independence of Irrelevent Alternatives. What they have done is to confuse this condition with another, unrelated condition. Here is Arrow's putative example of a voting procedure that violates Independence of Irrelevant Alternatives:

> [Consider] the rank-order method of voting frequently used in clubs. . . . let each individual rank all the candidates, i.e., designate his first-choice candidate, second-choice candidate, etc. Let preassigned weights be given to the first, second, etc., choices, the higher weight to the higher choice, and let the candidate with the highest weighted sum of votes be elected. In particular, suppose that there are three voters and four candidates, *x, y, z,* and *w*. Let the weights for the first, second, third, and fourth choices be 4, 3, 2, and 1, respectively. Suppose that individuals 1 and 2 rank the candidates in the order *x, y, z,* and *w,* while individual 3 ranks them in the order *z, w, x,* and *y*. Under the given electoral system, *x* is chosen, Then, certainly, if *y* is deleted from the ranks of the candidates, the system applied to the remaining candidates should yield the same result, especially since, in this case, *y* is inferior to *x*

according to the tastes of every individual; but, if y is in fact deleted, the indicated electoral system would yield a tie between x and z.

Arrow is flatly mistaken. In his example, there are two potential feasible sets, one obtained from the other by deleting y. The collective choice from each set depended only on voter preferences among members of *that set*, not among nonmembers, so Independence of Irrelevant Alternatives is satisfied. What is violated is a very different condition: that collective choices be independent of the feasibility of alternatives that would not have been chosen anyway—that if $y \in \alpha - C(\alpha)$ then $C(\alpha) = C(\alpha - \{y\})$. I discuss the latter condition in chapter 10. It is really a "rationality" condition, for it follows from BICH and P + I-Transitivity. It requires the collective choice to remain invariant when individual preferences are fixed and the feasible set varies a certain way. By contrast, Independence of Irrelevant Alternatives requires the collective choice to remain invariant when the *feasible set* is fixed and the *individual preferences* vary a certain way.

The standard case for the *desirability* of violating Binary Independence (and therewith Independence of Irrelevant Alternatives) rests on examples like this: The president of the United States has decided to settle a current dispute with Liechtenstein by diplomacy or to devastate Liechtenstein with a thermonuclear device but has not decided which. To reach a decision, the president convenes a committee consisting of herself, the secretary of state, and the chairman of the Joint Chiefs of Staff. The secretary votes for diplomacy, the chairman for devastation. Because her advisers are divided, she asks them to consider a third alternative, which she has already ruled out as impossible: impose economic sanctions on Liechtenstein. The two men rank the three alternatives in order of preference as follows:

Secretary Frank X. Changes	General Adam Bahm
Economic Sanctions	Devastation
Diplomacy	Economic sanctions
Devastation	Diplomacy

As a group, the advisers incline more toward devastation than toward diplomacy—or so the argument runs. Because the general ranked the infeasible alternative *between* the two feasible alternatives and the secretary did not, the general's preference for devastation over diplomacy seems *more intense* than the secretary's opposite preference.

This case against Binary Independence is weak for four reasons:

1. As I have argued already, it is questionable at best whether collective choices should reflect preference intensities, even were these known.

2. The preferences of the secretary and general depicted above give scant evidence for the hypothesis that the general's preference is more intense than the secretary's. For it could easily happen, in the example, that the general prefers devastation to sanctions very slightly and sanctions to diplomacy very slightly while the secretary prefers diplomacy to devastation very intensely.

3. The secretary should have no trouble finding another feasible alternative, say a chariot-led invasion of Liechtenstein across the Alps with catapult support, which he would rank between diplomacy and devastation but which the general would rank below diplomacy.

4. Even if no such fourth alternative can be found, this could be explained, not by the secretary's having a *less intense preference* than the general, but by the secretary's being *less sensitive*—less able to make subtle discriminations—than the general.

There are hypothetical examples in which preferences involving infeasible alternatives do provide some evidence for interpersonal comparisons of preference intensity. Suppose there are two feasible alternatives, x and y, which two individuals rank along with two infeasible alternatives as follows:

Mr. 1	Mr. 2
x + \$1 million	y + \$1
y + \$1	y
x	x + \$1 million
y	x

Then it is almost certain that Mr. 2's preference for y over x is more intense than Mr. 1's preference for x over y.

The trouble is, such extreme cases are rare, and they support only *numerically inspecific* comparisons—which do us no good when there are three or more individuals and we need to add "utility" numbers in order to compare aggregate preference intensities.

Another objection to Binary Independence gets rediscovered periodically: If, as required by Binary Independence, we ignore diplomacy in the Liechtenstein example when comparing sanctions with devastation, it seems sanctions and devastation should be collectively indifferent (neither collectively preferred to the other). And if we ignore sanctions when comparing devastation with diplomacy, it seems the latter should be collectively indifferent as well. At this point in the argument it is assumed that collective indifference should be transitive, whence it follows that sanctions and diplomacy should be collectively indifferent. But that is absurd: both men prefer sanctions to diplomacy.

The defect in this argument is the unsupported assumption that collective indifference should be transitive. If anything, the situation described is a counterexample to *this* assumption.

Besides ruling out one putative way of basing collective choices on preference intensities, does Binary Independence (or Independence of Irrelevant Alternatives) also rule out the use of preference intensities *as such*?

Yes, in one sense: it requires collective choices to depend only on such preferential information as could be got by taking a vote (as voting is customarily practiced).

And no, in another sense: it allows us to interpret $\mathbb{P}$ so that $x\mathbb{P}^v y$ holds when, in the v-situation, those who prefer x to y have a greater aggregate preference intensity than those who prefer y to x. For suppose $\mathbb{P}$ is interpreted that way. And suppose every $v \in PROF$ represents a situation with the following odd feature: All of Mr. 1's preferences are twice as intense as any of Mr. 2's, all of Mr. 2's are twice as intense as any of Mr. 3's, and so on. Then $x\mathbb{P}^v y$ holds if, and only if, either Mr. 1 prefers x to y, or Mr. 1 is indifferent between x and y and Mr. 2 prefers x to y, or Messrs. 1 and 2 are indifferent between x and y and Mr. 3 prefers x to y, or the like. But this obviously implies that $\mathbb{P}$ satisfies Binary Independence. Therefore, it is *possible* to interpret $\mathbb{P}$ and *PROF* so that $\mathbb{P}$ reflects preference intensities yet satisfies Binary Independence.

To show this, I imposed a severe and bizarre restriction on the range of situations represented by *PROF*, one that forced the maximization of aggregate preference satisfaction to coincide with "serial dictatorship" inside that range of situations. (As you will see in Part II, this restriction or something of its ilk was essential to the example.)

What the example proves is that serial dictatorship (which satisfies Binary Independence) and the maximization of aggregate preference satisfaction (which reflects preference intensities) might be equivalent (coincident) procedures within the range of situations represented by *PROF*, even if *PROF* satisfies Unrestricted Domain, although they obviously diverge in other possible situations—ones not represented by *PROF*.

2.4. SIMPLE MAJORITY RULE

Binary Independence requires that ℙ reflect no *preferential* information not obtainable by taking a vote (as voting is customarily practiced). It does *not* require that ℙ reflect no *information* not obtainable by taking a vote: ℙ might be a covert function of nonpreferential information of some sort. Suppose we further constrain the relevant information by requiring that ℙ reflect *nothing but voting* information—no information, preferential or nonpreferential, other than individual preferences between feasible alternatives. What does this requirement do to ℙ? It comes close to ensuring that ℙ represent *Simple Majority Rule.* It does ensure this once we add the positive requirement that ℙ be sensitive to all available voting information (not just insensitive to all other information).

Simple Majority Rule (in the sense in which I shall use this phrase) is a collective-choice process applicable only to two-element feasible sets. In a contest between two alternatives, Simple Majority Rule prescribes the choice of whichever alternative has the greater number of supporters, allowing the choice of either alternative if their supporters are equally numerous.

There are well-known ways to generalize Simple Majority Rule to the multialternative case—Plurality Rule, Simple Runoff Rule, Borda (Point Count) Rule, and so forth. Each has its partisans, each its critics. I discuss these voting procedures in chapter 8. For now I am concerned only with pairwise collective choices.

Simple Majority Rule holds a distinguished position among collective-choice processes. Barring special circumstances, when a collective choice is to be made between two alternatives, Simple Majority Rule almost invariably is used; we tend instinctively to identify pairwise collective choice with majority choice, and political democracy with majoritarianism. Why? Perhaps because Simple Majority Rule is the one and only two-alternative collective-choice process that is sensitive to available voting information but reflects *no other information*.

To make this precise, assume Axioms $\mathbb{P}$ and *PROF*, and assume that every possible way Messrs. $1, 2, \ldots, n$ can vote or abstain in a contest between any two given members of A is represented by some preference profile in *PROF*:

> STRONG 2-DOMAIN *For all distinct $x, y \in A$ and all disjoint subsets K, M of N, there is a $v \in PROF$ such that $K = \{i \mid xP_i^v y\}$ and $M = \{i \mid yP_i^v x\}$.*

This differs from 2-Domain in that it requires *PROF* to contain profiles in which some individuals are indifferent between two alternatives.

Suppose the two-alternative collective-choice process represented by $\mathbb{P}$ is sensitive to all available information about the way Messrs. $1, 2, \ldots, n$ vote (or *would* vote) but is independent of all other information.

Here is a reasonable way to make the sensitivity part of this assumption more precise: If, on the basis of available information, the process prescribes neither the choice of x nor that of y in a contest between the two, and if further information is then obtained that unquestionably weighs in favor of x, that is enough to tilt the balance, causing the process to prescribe the choice of x. Formally:

> SENSITIVITY *Suppose x, y are distinct elements of A, $v, w \in PROF$, and $x\mathbb{I}^v y$. And suppose $\{i \mid yP_i^v x\} = \{i \mid yP_i^w x\}$ while $\{i \mid xP_i^v y\} \subset \{i \mid xP_i^w y\}$. Then $x\mathbb{P}^w y$.*

In other words: If, faced with a deadlock between x and y, some erstwhile abstainers come round and vote for x, other things remaining the same, that is enough to break the deadlock in x's favor.

Now suppose the process in question prescribes the choice of x in preference to y. Because the only properties of x and y that are relevant to this choice consist of the way Messrs. $1, 2, \ldots, n$ voted in the x-vs.-y contest, if we replace x and y in everyone's ballot by other alternatives,

x' and y', then the latter will have the same relevant properties x and y had, and so the process will prescribe the choice of x' in preference to y'. Formally:

NEUTRALITY *Suppose $v, v' \in PROF$, $x, y, x', y' \in A$, and for all $i \in N$, $x\mathbb{P}_i^v y$ iff $x'P_i^{v'}y'$ and $yP_i^v x$ iff $y'P_i^{v'}x'$. Then if $x\mathbb{P}^v y$, $x'\mathbb{P}^{v'}y'$.*

Suppose again the process in question prescribes the choice of x in preference to y. Because this choice depends on no properties of Messrs. 1, 2, . . . , n other than the way they voted, if two of these individuals switch votes, each voting the way the other did, that cannot alter the collective choice: the process will still prescribe the choice of x. Similarly if many such switches take place. Formally:

ANONYMITY *Suppose $v, v' \in PROF$, $x, y \in A$, and f is a permutation of N such that, for all $i \in N$, $xP_i^v y$ iff $xP_{f(i)}^{v'}y$ and $yP_i^v x$ iff $yP_{f(i)}^{v'}x$. Then if $x\mathbb{P}^v y$, $x\mathbb{P}^{v'}y$.*

The "one-man, one-vote" doctrine formalized by Anonymity is the sort of thing on behalf of which people take to the barricades. Yet within a more or less democratic context, it is rather Neutrality that is the more important, more interesting, more mooted condition. Take the Two-Thirds Majority Rule used in the U.S. Senate to ratify treaties. Whenever the rule is used, there are just two alternatives, ratification and rejection. They are not treated equally, contrary to Neutrality: rejection has a two-to-one edge over ratification. More generally, so-called special-majority (as opposed to Simple Majority) rules always give an advantage to "negative" or status quo alternatives.

Given the assumptions I have made, $\mathbb{P}$ must represent Simple Majority Rule; that is, $\mathbb{P}$ must satisfy:

SIMPLE MAJORITY CONDITION (SMAC) *For all $v \in PROF$, $x\mathbb{P}^v y$ iff $|\{i \mid xP_i^v y\}| > |\{i \mid yP_i^v x\}|$.*

In fact, assuming Axioms *PROF* and $\mathbb{P}$ and Strong 2-Domain, SMAC is *equivalent* to the conjunction of Sensitivity, Neutrality, and Anonymity. So Simple Majority Rule is the one and only two-alternative collective-choice process meeting these conditions.

That SMAC *implies* the three conditions is obvious. To prove the converse, I need two lemmata.

LEMMA 2.4.1. *Assume Axioms PROF and* P, *Strong 2-Domain, Neutrality, and Anonymity. Suppose* $v \in PROF$, $x, y \in A$, *and* $|\{i \mid xP_i^v y\}| = |\{i \mid yP_i^v x\}|$. *Then not* $x\mathrm{P}^v y$.

Proof. Suppose $x\mathrm{P}^v y$; to deduce a contradiction. Let

$$K = \{i \mid xP_i^v y\} \qquad \text{and} \qquad M = \{i \mid yP_i^v x\}.$$

Since K and M have the same cardinality, there is a one-to-one mapping e of K onto M. Define the permutation f of N this way:

$$f(i) = \begin{cases} e(i) & \text{when } i \in K, \\ e^{-1}(i) & \text{when } i \in M, \\ i & \text{when } i \in N - (K \cup M). \end{cases}$$

Because each P_i^v is asymmetric, K and M are disjoint. And since $x\mathrm{P}^v y$, x and y are distinct members of A. So by virtue of Strong 2-Domain, there is a $v' \in PROF$ for which

$$M = \{i \mid xP_i^{v'} y\} \qquad \text{and} \qquad K = \{i \mid y\, P_i^{v'} x\}.$$

But then, since $K = \{i \mid xP_i^v y\}$ and M is the image of K under f,

$$xP_i^v y \text{ iff } xP_{f(i)}^{v'} y, \qquad i = 2, \ldots, n.$$

And since $M = \{i \mid yP_i^v x\}$ and K is the image of M under f,

$$yP_i^v x \text{ iff } yP_{f(i)}^{v'} x, \qquad i = 1, 2, \ldots, n.$$

Hence, by Anonymity, $x\mathrm{P}^{v'} y$. Because

$$K = \{i \mid xP_i^v y\} = \{i \mid yP_i^{v'} x\}$$

and

$$M = \{i \mid yP_i^v x\} = \{i \mid xP_i^{v'} y\},$$

we have, for all $i \in N$,

$$xP_i^{v'} y \text{ iff } yP_i^v x$$

and

$$yP_i^{v'} x \text{ iff } xP_i^v y.$$

Therefore, since $x\mathrm{P}^{v'} y$, Neutrality implies that $y\mathrm{P}^v x$. But in view of the asymmetry of P^v, that contradicts the assumption that $x\mathrm{P}^v y$. ■

LEMMA 2.4.2. *Assume Axioms PROF and* $\mathbb{P}$, *Strong 2-Domain, Sensitivity, Neutrality, and Anonymity. Then SMAC holds. That is,* $x, y \in A$, and $|\{i|xP_i^vy\}| > |\{i|yP_i^vx\}|$. *Then* $x\mathbb{P}^vy$.

Proof. Let $M = \{i|yP_i^vx\}$, and let K be a proper subset of $\{i|xP_i^vy\}$ with the same cardinality as M. By hypothesis, for some i, xP_i^vy, whence $x \neq y$. So by virtue of Strong 2-Domain, there is a $w \in PROF$ such that

$$K = \{i|xP_i^wy\} \qquad \text{and} \qquad M = \{i|yP_i^wx\}.$$

By Lemma 2.4.1, then, since K has the same cardinality as M, neither $x\mathbb{P}^wy$ nor $y\mathbb{P}^wx$. But

$$M = \{i|yP_i^wx\} = \{i|yP_i^vx\}$$

and

$$K = \{i|xP_i^wy\} \subset \{i|xP_i^vy\}$$

Hence, by Sensitivity, $x\mathbb{P}^vy$. ■

THEOREM 2.4.1. *Assume Axioms PROF and* $\mathbb{P}$, *Strong 2-Domain, Sensitivity, Neutrality, and Anonymity. Then SMAC holds. That is, for all* $v \in PROF$, $x\mathbb{P}^vy$ *iff* $|\{i|xP_i^vy\}| > |i|yP_i^vx\}|$.

Proof. Trivial if x, y are not distinct elements of A. Otherwise, if $K = \{i|xP_i^vy\}$ is larger than $M = \{i|yP_i^vx\}$ then $x\mathbb{P}^vy$ by Lemma 2.4.2; and if K is not larger than M then either $|K| = |M|$, whence not $x\mathbb{P}^vy$ by Lemma 2.4.1, or else $|M| > |K|$, whence $y\mathbb{P}^vx$ by Lemma 2.4.2, and thus not $x\mathbb{P}^vy$. ■

SOURCES AND RELATED CONTRIBUTIONS

The choice-function concept introduced in chapter 1 apparently originated with Arrow (1952 and 1963: ch. 2). It generalizes the economist's concept of a demand function as well as the set theorist's concept of a choice function. Both demand functions and set-theoretic choice functions yield single elements rather than subsets of the sets to which they are applied. Because that feature is essential to choice functions in the orthodox, set-theoretic sense—the sense relevant to the Axiom of

Choice—I resisted the Arrovian usage for several years (see, for example, Schwartz 1972), surrendering when I found I had no troops.

My terminology for ordering relations is traditional (see Suppes 1960: 68–72). The semiorder concept comes from Luce (1956), reflecting an idea of Armstrong (1939). The more general interval-order concept is due to Fishburn (1970c). Because of the large number of transitivity conditions discussed in Part I and later, I have adopted mnemonic names for them. Sen (1970: ch. 1 and 1*) christens two of these conditions differently: my *P + I-Transitivity* is his *Transitivity*, and my *P-Transitivity* is his *Quasi-Transitivity*. For the representation of ordering relations by real-valued "utility" functions, see Debreu (1954; 1959:7–9), Luce (1956), and Fishburn (1970c,d; 1972).

For the classical "rationality" conditions on choice functions, see Arrow (1952 and 1963: ch. 2), Debreu (1959:7–9), Samuelson (1972), and Schwartz (1972, 1976). These conditions formulate the widely accepted equation of *rational* choice with *maximizing* choice.

I have left individual preferences and modes of individual participation in collective-choice processes unanalyzed. It has long been recognized that an individual might vote (or otherwise express his preferences) either *sincerely* or *strategically* (or, I should add, in any number of other ways, such as symbolically). A strategic voter expresses something other than his true preferences in order thereby to secure a collective choice that he prefers to the one which would have prevailed had he expressed his true preference. For a well-known treatment of strategic voting, see Farquharson (1969). An intensively investigated question is whether there can be a collective-choice process that is *strategy proof* in the sense that it provides no incentive for individuals to vote strategically. Given some mild assumptions, Gibbard (1973) and Satterthwaite (1975) have proved that no *resolute* collective-choice process—no collective-choice process that yields only single-member choice sets —can be strategy proof. Schwartz (1982b) has proved a similar result without the resoluteness restriction, a restriction satisfied by no familiar voting rule. It remains open whether collective-choice processes can be strategy proof for *reasonably restricted domains* of preference profiles. See also the important contributions of Gibbard (1977), Kelley (1977), and Pattanaik (1978).

My social-preference function $\mathbb{P}$ differs from an Arrovian *social welfare function* (Arrow 1950, 1952, 1963) in two respects: First, the values of a social welfare function must, by definition, be weak orderings, but the values of ℙ need not be. Second, like Fishburn (1973) but unlike Arrow, I represent individual and collective preference (the arguments and values of ℙ) by asymmetric relations (relations of *strict* preference) rather than reflexive relations (relations of preference or indifference). The second difference is merely stylistic: a relation of either sort can be defined in terms of a relation of the other sort.

On measuring preference intensities and making interpersonal comparisons, see Little (1957), Arrow (1963:112–120; 1967b, 1977b), Luce and Raiffa (1957), Rothenberg (1961: chs. 7 and 8), Goodman and Markowitz (1952), Harsanyi (1955), Churchman (1966), Sen (1970: chs. 7, 7*, 8, and 8*), Waldner (1972), MacKay (1975), Schwartz (1975a), Strasnick (1976), and d'Aspremont and Gevers (1977). See also section 4.3.

The Independence of Irrelevant Alternatives condition comes from Arrow (1950, 1952, 1963). So, of course, does the mistaken application of it. Luce and Raiffa (1957: ch. 14) repeat the mistake. Rothenberg (1961: ch. 6) handles the condition better than many subsequent treatments. More than anyone else, Plott (esp. 1971, 1972, 1976) has explained the condition and worked to expose erroneous interpretations. Arrow's mistake was first pointed out by Plott (1971: 116 and Note 13). Hansson (1973) ingeneously defends a view contrary to Plott's and mine. See also Ray (1973), Osborne (1976), and Karni and Schmiedler (1976). The arguments against Independence come from Luce and Raiffa. The lovely example involving y + \$1 and x + \$1,000,000 was given me by Martin Bailey.

On the connection between preference and welfare in a single subject, see Barry (1965: chs. 3, 10, 11, and 12), Braybrooke (1968), Arrow (1967b), Brandt (1967), Schwartz (1973: ch. 3), and Galston (1980). In Schwartz (1982a) I extend the discussion above, arguing that welfare is a nonsubjective concept : what is good for a person has little to do with his preferences.

Theorem 2.4.1, which gives an axiomatic characterization of Simple Majority Rule, comes essentially from May (1952, 1953). Because May deals with a single two-alternative choice, he can avoid my general framework and give a beautifully succinct formulation: Letting +1 represent one alternative, −1 the other, and 0 indifference between these alternatives, a *profile* is a point in $\{0, 1, -1\}^n$, and a collective-choice rule is a function $f:\{0, 1, -1\}^n \rightarrow \{0, 1, -1\}$. In this framework, Anonymity says that f is a symmetric function ($f(x_1, \ldots, x_i, \ldots, x_j, \ldots, x_n) \equiv f(x_1, \ldots x_j, \ldots, x_i, \ldots, x_n)$) and Neutrality says that f is a negative function ($f(-x_1, \ldots, -x_n) \equiv -f(x_1, \ldots, x_n)$). Murakami (1968) and Fishburn (1973: chs. 2–6) generalize May's approach from Simple Majority Rule to pairwise collective-choice processes of all sorts. Murakami draws upon many-valued logic while Fishburn uses linear algebra.

PART II

COLLECTIVE CHOICE AND COLLECTIVE RATIONALITY

CHAPTER THREE

VOTING PARADOXES

3.1. THE CLASSICAL VOTING PARADOX: THEME AND VARIATIONS

Intuitively appealing though they be, the classical "rationality" conditions on collective-choice processes conflict with some apparently reasonable and realistic conditions that relate collective choices to individual preferences. Results to this effect are called "impossibility theorems" and regarded as paradoxes. The first two were the Classical Voting Paradox, discovered by the Marquis de Condorcet at the end of the eighteenth century and presented in this section, and Arrow's Paradox (or Impossibility Theorem), discovered by Kenneth Arrow in the middle of this century and presented in the next section. Most of this book is an attempt to generalize and interpret these two paradoxes and to apply their lessons. The chief lesson, I will argue, is that the traditional "rationality" conditions are unacceptable from both a positive and a normative point of view.

A virtue of BICH is that it reduces the problem of making, explaining, and predicting choices generally to that of making, explaining, and predicting *pairwise* choices. Its great drawback is that it implies P-Acyclicity (assuming Axiom C1), a most severe restriction on pairwise choices.

The severity of this restriction is brought out by the Classical Voting Paradox: Let three voters rank three alternatives in order of preference as follows:

Mr. 1	Mr. 2	Mr. 3
x	y	z
y	z	x
z	x	y

A majority (Messrs. 1 and 3) prefer x to y, a majority (Messrs. 1 and 2) prefer y to z, and a majority (Messrs. 2 and 3) prefer z to x. So if P is the relation of majority preference then $xPyPzPx$, contrary to P-Acyclicity and therewith BICH. To put the problem another way: With P interpreted as majority preference, BICH says $C(\{x, y, z\})$ comprises those members of $\{x, y, z\}$ that are not beaten, under Simple Majority Rule, by anything in $\{x, y, z\}$. The trouble is that everything in $\{x, y, z\}$ is beaten by something in $\{x, y, z\}$.

Let me restate this problem more formally and generally. What we have seen, in effect, is that these four conditions are inconsistent:

$n = 3$, 3-Domain, SMAC,

COLLECTIVE P-ACYCLICITY *For no* $v \in PROF$ *is there a* P^v*-cycle* (P^v *is a suborder for all* $v \notin PROF$).

As it happens, the assumption that $n = 3$ is inessential, and SMAC is unnecessarily strong. There are so-called cyclic majorities involving any number of voters greater than four, and the inconsistency remains when SMAC is weakened as follows:

ABSOLUTE MAJORITY CONDITION *If* $|\{i \mid xP_i^v y\}| > n/2$ then $xP^v y$,

THEOREM 3.1.1 *If* $n = 3$ *or* $n > 4$, *these three conditions are inconsistent: 3-Domain, Absolute Majority Condition, and Collective P-Acyclicity.*

Proof. Assume the conditions; to deduce a contradiction. By 3-Domain, A has at least three distinct elements, x, y, and z.

Suppose $n = 3$. By 3-Domain again, there is a $v \in PROF$ whose component relations, P_1^v, P_2^v, and P_3^v, linearly order $\{x, y, z\}$ as follows:

1	2	3
x	y	z
y	z	x
z	x	y

By Absolute Majority Condition, $x\mathrm{P}^v y\mathrm{P}^v z\mathrm{P}^v x$, contrary to Collective P-Acyclicity.

Now suppose $n > 4$. Let

m = the least integer greater than $n/2$,
$M^x_y = \{1, 2, \ldots, m\}$,
$M^y_z = \{m-1, m, m+1, \ldots, n\}$, and
$M^z_x = N - \{m-1, m\} = N - (M^x_y \cap M^y_z)$.

Then M^x_y and M^y_z have more than $n/2$ members each. And since $n > 4$ and M^z_x has $n-2$ members, M^z_x has more than $n/2$ members as well. By 3-Domain, there is a $v \in PROF$ whose component relations linearly order $\{x, y, z\}$ as follows:

M^x_y (over $1, \ldots, m$)							
M^z_x (over $1, \ldots, m-2$)					M^z_x (over $m+1, \ldots, n$)		
1	. . .	$m-2$	$m-1$	m	$m+1$	. . .	n
z		z					
x	. . .	x	x	x			
y		y	y	y	y		y
			z	z	z	. . .	z
					x		x
			M^y_z (under $m-1, \ldots, n$)				

Then $M^x_y = \{i \mid xP^v_i y\}$, $M^y_z = \{i \mid yP^v_i z\}$, and $M^z_x = \{i \mid zP^v_i x\}$. Thus, since M^x_y, M^y_z, and M^z_x have more than $n/2$ members each, Absolute Majority Condition implies that $x\mathrm{P}^v y\mathrm{P}^v z\mathrm{P}^v x$, contrary to Collective P-Acyclicity. ■

The reason for assuming $n \neq 4$ in Theorem 3.1.1 is that there can be no three-alternative majority-preference cycle corresponding to a profile of four linear preferences (try constructing one).

The phenomenon of cyclic collective preference is not peculiar to Simple Majority Rule. There are possible situations in which Collective P-Acyclicity conflicts with the special-majority rules (Two-Thirds, Three-Quarters, and the like), and indeed with any two-alternative collective-choice rule that prescribes the choice of one alternative over another whenever all but at most a single individual prefer the former to the latter. For the inconsistency remains when we replace 3-Domain by n-Domain and further replace Absolute Majority Condition by:

VIRTUAL UNANIMITY *If* $xP^v_{N-\{k\}}y$ *then* $x\mathbb{P}^v y$.

THEOREM 3.1.2. *n-Domain, Virtual Unanimity, and Collective P-Acyclicity are jointly inconsistent.*

Proof. Assume the three conditions; to deduce a contradiction. By n-Domain, there are $x_1, \ldots, x_n \in A$ and a $v \in PROF$ such that, for every $i \in N$, P^v_i linearly orders $\{x_1, \ldots, x_n\}$ s follows:

$$x_i, x_{i+1}, \ldots, x_n, x_1, x_2, \ldots, x_{i-1}.$$

For each $i \neq 1$ in N, x_n precedes x_1 in the P^v_i-ordering of $\{x_1, \ldots, x_n\}$. That is, $x_n P^v_i x_1$ for each $i \in N - \{1\}$. Hence, by Virtual Unanimity,

$$x_n\, \mathbb{P}^v x_1.$$

Now let $k \in N - \{n\}$. Then for each $i \in N$, x_k precedes x_{k+1} in the P^v_i-ordering of $\{x_1, \ldots, x_n\}$ unless x_k is the last element, x_{i-1}, in that ordering—unless $i = k + 1$, that is. In short, $x_k P^v_i x_{k+1}$ for each $i \neq k + 1$ in N. Hence, by Virtual Unanimity, $x_k \mathbb{P}^v x_{k+1}$. Consequently, since our choice of k was arbitrary,

$$x_1 \mathbb{P}^v x_2 \mathbb{P}^v x_3 \cdots x_{n-1} \mathbb{P}^v x_n \mathbb{P}^v x_1,$$

contrary to Collective P-Acyclicity. ■

The proofs of Theorems 3.1.1 and 3.1.2 generalize Condorcet's three-man, three-alternative example in different ways. Condorcet's example is the three-man case of *two minimally overlapping bare majorities*, the construction used to prove Theorem 3.1.1. That example also is a three-by-three *latin square*—a three-by-three array of the following general form:

$$\begin{array}{cccccccc}
x_1 & x_2 & \cdots & x_i & & \cdots & x_n \\
x_2 & x_3 & \cdots & x_{i+1} & & \cdots & x_1 \\
\vdots & \vdots & & \vdots & & & \vdots \\
 & & & x_n & & & \\
 & & & x_1 & & & \\
 & & & \vdots & & & \\
 & x_n & & & & & \\
x_n & x_1 & \cdots & x_{i-1} & & \cdots & x_{n-1}
\end{array}$$

And this is the form of profile used in the proof of Theorem 3.1.2.

The immediate lesson of these results is that Majority Rule and some other pairwise collective-choice processes can (depending on individual preferences) violate P-Acyclicity and therewith BICH. Before attempting to interpret this fact, let us expand our "data base" by examining Arrow's Paradox and some variations on the Arrovian theme.

3.2. ARROW'S PARADOX

Theorem and Proof

Kenneth Arrow's celebrated Impossibility Theorem asserts the inconsistency of the following seven conditions on a pairwise collective-choice process:

AXIOM $\mathbb{P}$ *$\mathbb{P}$ is a function on PROF whose value, $\mathbb{P}^v$, at each $v \in PROF$ is an asymmetric binary relation on A.*

UNRESTRICTED DOMAIN (a) *Every vector in PROF comprises only weak orderings of A.* (b) *A is completely free, that is, every ordered n-tuple of weak orderings of A belongs to PROF.*

$|A| \geq 3$.

BINARY INDEPENDENCE *If $v, w \in PROF$, $v/\{x, y\} = w/\{x, y\}$, and $x\mathbb{P}^v y$, then $x\mathbb{P}^w y$.*

PARETO PREFERENCE *If $xP^v_N y$ then $x\mathbb{P}^v y$.*

NONDICTATORSHIP *There is no i such that $\{i\}$ is superdecisive.*

COLLECTIVE P + I-TRANSITIVITY *If $x\mathbb{P}^v y\mathbb{P}^v z$ then $x\mathbb{P}^v z$, and if $x\mathbb{I}^v y\mathbb{I}^v z$ then $x\mathbb{I}^v z$ ($\mathbb{P}^v$ is a [strict] weak ordering of A for all $v \in PROF$).*

Although itemized, labeled, and worded a bit differently, these conditions clearly are Arrow's (1952, rev. 1963). I have already discussed Axiom $\mathbb{P}$, Unrestricted Domain, and Binary Independence. According to Pareto Preference, every unanimous preference is a collective preference: if Messrs. 1, 2, . . . , n all prefer x to y then x is collectively preferred to y. Nondictatorship says no one individual has so much power that he can dictate the choice between *any* two alternatives belonging to A in *every* possible situation represented by *PROF*—including all those possible situations in which he is unanimously opposed. Collective P + I-Transitivity just says that the collective-preference relation always satisfies P + I-Transitivity.

Although Arrow expressly assumed Unrestricted Domain, all he actually used in his proof was the far less demanding 3-Domain. In my statement and proof of Arrow's theorem, I replace Unrestricted Domain by 3-Domain—which implies that $|A| \geq 3$, eliminating the need for a separate assumption to that effect.

To prove Arrow's Impossibility Theorem, I need Theorem 2.3.1 plus three lemmata. Note that the first two do not depend on the full force of Collective P + I-Transitivity, but only the weaker:

COLLECTIVE P-TRANSITIVITY If xP^vyP^vz *then* xP^vz(P^v *is a strict partial order for all* $v \in PROF$).

LEMMA 3.2.1. *Assume 3-Domain, Binary Independence, Pareto Preference, and Collective P-Transitivity. Suppose g is decisive for x vs. y. Then:*
(a) *g is decisive for x vs. every* $z \in A - \{x\}$.
(b) *g is decisive for every* $z \in A - \{y\}$ *vs.* y.
(c) *g is decisive for every* $z \in A$ *vs. every* $w \in A - \{z\}$.

Proof. (a) Trivial if $z = y$. Otherwise, by 3-Domain, there is a $v \in PROF$ whose component relations order $\{x, y, z\}$ as follows:

g	$N-g$
x	y
y	z
z	x

xP^vy by the decisiveness of g, and yP^vz by Pareto Preference, whence xP^vz by Collective P-Transitivity, so g is decisive for x vs. z by Theorem 2.3.1.

(b) Trivial if $z = x$. Otherwise, by 3-Domain, some $v \in PROF$ has this form:

g	$N-g$
z	y
x	z
y	x

zP^vx by Pareto Preference, and xP^vy by the decisiveness of g, whence zP^vy by Collective P-Transitivity, so g is decisive for z vs. y by Theorem 2.3.1.

(c) By 3-Domain, let $t \in A - \{x, z\}$. Then by hypothesis and (a),

	g is decisive for x vs. t,	
whence	g is decisive for z vs. t	by (b),
so	g is decisive for z vs. w	by (a). ■

LEMMA 3.2.2. *Assume 3-Domain, Binary Independence, Pareto Preference, and Collective P-Transitivity. Suppose g is decisive for some pair. Then g is superdecisive; that is, the following two conditions hold for all distinct* $x, y \in A$:

(a) *Every h for which* $g \subseteq h \subseteq N$ *is decisive for x vs. y.*

(b) *Assuming every three-element subset of A is completely free, then for every* $v \in PROF$, xP^vy *if* xP^v_gy.

Proof. By hypothesis and Lemma 3.2.1, g is decisive for every pair of distinct elements of A.

To prove (a), suppose $g \subseteq h \subseteq N$. By hypothesis and 3-Domain, there is a $z \in A - \{x, y\}$ and a $v \in PROF$ of this form:

h		
g	$h-g$	$N-h$
x	z	z
z	x	y
y	y	x

xP^vz by the decisiveness of g, and zP^vy by Pareto Preference, whence xP^vy by Collective P-Transitivity, so h is decisive for x vs. y by Theorem 2.3.1.

To prove (b), suppose every three-element subset of A is completely free, and xP^v_gy. By 3-Domain, let $z \in A - \{x, y\}$. Then $\{x, y, z\}$ is completely free. So some $w \in PROF$ has this form:

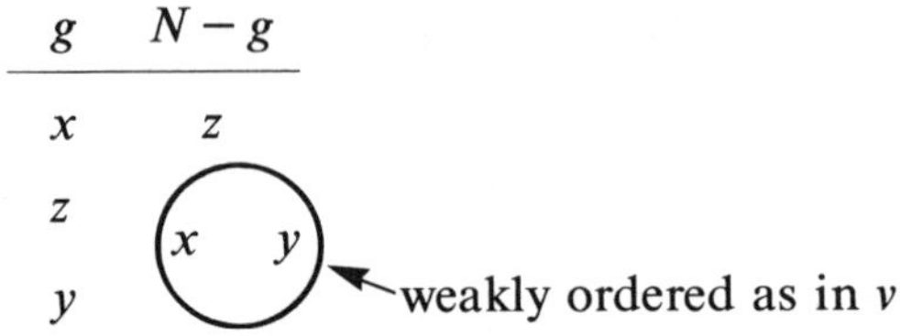

$x\mathbb{P}^w z$ by the decisiveness of g, and $z\mathbb{P}^w y$ by Pareto Preference, whence $x\mathbb{P}^w y$ by Collective P-Transitivity. But $v/\{x, y\} = w/\{x, y\}$. So $x\mathbb{P}^v y$ by Binary Independence. ■

LEMMA 3.2.3. *Assume Axiom* $\mathbb{P}$ *and Collective P + I-Transitivity. Suppose* $x\mathbb{P}^v y$, $z \in A$, *and not* $z\mathbb{P}^v y$. *Then* $x\mathbb{P}^v z$.

Proof. Suppose *not* $x\mathbb{P}^v z$; to deduce a contradiction. If $y\mathbb{P}^v z$ then $x\mathbb{P}^v z$ by hypothesis and Collective P + I-Transitivity. Hence,

not $y\mathbb{P}^v z$.

And if $z\mathbb{P}^v x$ then $z\mathbb{P}^v y$ by hypothesis and Collective P + I-Transitivity, contrary to hypothesis. Hence,

not $z\mathbb{P}^v x$.

By Axiom $\mathbb{P}$, since $z \in$ A and $x\mathbb{P}^v y$, we have $v \in$ *PROF* and $x, y, z \in A$. Thus, since not $z\mathbb{P}^v y$, not $x\mathbb{P}^v z$, not $y\mathbb{P}^v z$, and not $z\mathbb{P}^v x$, the definition of $\mathbb{I}$ implies

$x\mathbb{I}^v z$ and $z\mathbb{I}^v y$.

By Collective P + I-Transitivity, then, $x\mathbb{I}^v y$, whence not $x\mathbb{P}^v y$, contrary to hypothesis. ■

THEOREM 3.2.1 (*Arrow's Impossibility Theorem*). *The following six conditions are jointly inconsistent:*

Axiom $\mathbb{P}$,	*3-Domain,*
Binary Independence,	*Pareto Preference,*
Nondictatorship, and	*Collective P + I-Transitivity.*

(*In other words, there do not exist an integer n, sets A and PROF, and a function* $\mathbb{P}$ *satisfying all six conditions.*)

Proof. Assume the six conditions; to deduce a contradiction. By 3-Domain, A has at least two members, say t and u. So N is decisive for t vs. u by Pareto Preference, and thus some subset g of N is *minimally*

decisive in this sense: g is decisive for some pair, say x vs. y, and no smaller set is decisive for any pair. By Pareto Preference, $g \neq \emptyset$; say $i \in g$. By 3-Domain, there is a $z \in A - \{x, y\}$ and a $v \in PROF$ of this form:

g		
$\{i\}$	$g - \{i\}$	$N - g$
x	z	y
y	x	z
z	y	x

Because g is *minimally* decisive, $g - \{i\}$ cannot be decisive for z vs. y, and thus, by Theorem 2.3.1, not $z\mathrm{P}^v y$. But $x\mathrm{P}^v y$ by the decisiveness of g. So $x\mathrm{P}^v z$ by Lemma 3.2.3. By Theorem 2.3.1, then, $\{i\}$ is decisive for x vs. z. Hence, by Lemma 3.2.2, $\{i\}$ is superdecisive, contrary to Nondictatorship. ■

Lemma 3.2.2, along with the definition of superdecisiveness, could have been simplified had I joined Arrow in assuming the complete freedom of every three-element subset of A. And if Nondictatorship were replaced by the slightly stronger condition that no individual is decisive for every pair of distinct elements of A (rather than that no individual is superdecisive), we should not have needed Lemma 3.2.2 at all.

Weaker Rationality Conditions

Arrow's Theorem sometimes is regarded as a generalization of the Classical Voting Paradox. It is no such thing. True, Binary Independence, Pareto Preference, and Nondictatorship constitute a generalization of SMAC: SMAC implies them but does not follow from them; taken together, they are weaker than SMAC. But Collective P + I-Transitivity is *not* a generalization of Collective P-Acyclicity. On the contrary, Collective P-Acyclicity is a generalization of Collective P + I-Transitivity: Collective P + I-Transitivity implies Collective P-Acyclicity but not conversely. So Arrow's Paradox, although more general than the Classical Voting Paradox (or Theorem 3.1.1) in one respect, actually is *less* general in another.

This suggests exploring what happens when we weaken Collective

P + I-Transitivity. We might replace this condition by the weaker Collective P-Transitivity, for example, or by the still weaker Collective P-Acyclicity. Other possibilities include the various transitivity conditions stronger than Collective P-Transitivity but weaker than Collective P + I-Transitivity, such as:

COLLECTIVE PIP-TRANSITIVITY *If* $x\mathbb{P}^v y\mathbb{I}^v z\mathbb{P}^v w$ *then* $x\mathbb{P}^v w$ ($\mathbb{P}^v$ *is an interval order of A for all* $v \in PROF$).

and

COLLECTIVE PIP + IPP-TRANSITIVITY *If* $x\mathbb{P}^v y\mathbb{I}^v z\mathbb{P}^v w$ *then* $x\mathbb{P}^v w$, *and if* $x\mathbb{I}^v y\mathbb{P}^v z\mathbb{P}^v w$ *then* $x\mathbb{P}^v w$ ($\mathbb{P}^v$ *is a semiorder of A for all* $v \in PROF$).

Even Collective P-Acyclicity can be weakened in interesting ways. For example, Theorem 3.1.1 did not use the full force of Collective P-Acyclicity, but only the weaker:

COLLECTIVE P-ATRICYCLICITY *Not* $x\mathbb{P}^v y\mathbb{P}^v z\mathbb{P}^v x$ (*for no* $v \in PROF$ *is there a three-member* $\mathbb{P}^v$*-cycle*).

The most important of these conditions are Collective P-Acyclicity and weaker variants thereof. Because no rationality or transitivity condition stronger than Collective P-Acyclicity was involved in the Classical Voting Paradox (or in Theorems 3.1.1 and 3.1.2), no such stronger condition can yield a true generalization of the Classical Voting Paradox. And only when, for some $v \in PROF$, Collective P-Acyclicity is violated—only when there is a $\mathbb{P}^v$-cycle—is it impossible to make a $\mathbb{P}^v$-undominated collective choice—one to which no feasible alternative is collectively preferred.

If, in Arrow's set of conditions, we replace Collective P + I-Transitivity by *any* of the transitivity conditions just listed, we get a consistent set of conditions. To prove this, let

$$n \geq 2, \qquad A = \{1, 2, 3\}, \qquad \mathbb{P}^v \equiv P^v_N,$$

$$PROF = \{(\rho_1, \ldots, \rho_n) \mid \rho_1, \ldots, \rho_n \text{ are weak orderings of } A\}.$$

So A has just three members, and $\mathbb{P}^v$ is the relation of unanimous preference on A. It is obvious that Axiom $\mathbb{P}$, Unrestricted Domain (and therewith 3-Domain), Binary Independence, Pareto Preference, and Nondictatorship are satisfied. So is Collective P-Transitivity: if the preferences of Messrs. 1, 2, . . . , n are transitive, so must be their unanimous preference. Even the comparatively strong Collective PIP + IPP-Transitivity is satisfied, since in a three-element domain this condition is equivalent to Collective P-Transitivity.

Because Arrow's conditions are inconsistent, Collective P + I-Transitivity must be violated under the proposed interpretation. To see just why, consider a profile v of this form:

1	$n - \{1\}$
x	y
z	x
y	z

Because neither $xP^v_N y$ nor $yP^v_N x$, neither $x\mathbb{P}^v y$ nor $y\mathbb{P}^v x$, so that $x\mathbb{I}^v y$. Similarly, $y\mathbb{I}^v z$. According to Collective P + I-Transitivity, we should have $x\mathbb{I}^v z$. But in fact we have $x\mathbb{P}^v z$ because $xP^v_N z$.

For the interpretation to satisfy Collective P-Transitivity, it was not necessary that $|A| = 3$. This was necessary, however, in the case of Collective PIP-Transitivity and its ilk, as you will see in section 3.4.

Although Arrow's conditions become consistent when Collective P + I-Transitivity is weakened in any of the standard ways, the inconsistency recurs when certain modifications are made in his other conditions. That is, there are Arrow-style impossibility theorems involving each of the weaker rationality conditions lately listed, as well as others, as I shall now show.

3.3. COLLECTIVE P-TRANSITIVITY AND OLIGARCHIES

Take Arrow's conditions, replace Collective P + I-Transitivity by the weaker Collective P-Transitivity, and you have a consistent set of conditions. Still, the choice processes that satisfy these conditions are not very appealing. Although not dictatorial, they are *oligarchic.* Under any such process, there exists a group g—an oligarchy—with these two features:

(1) g is superdecisive.
(2) Assuming no one is indifferent between two given alternatives, neither alternative can be collectively preferred to the other unless *all* members of g prefer the one to the other; and this

feature extends to situations in which some members of *N-g* are indifferent between the two alternatives, so long as every three-element subset of *A* is completely free.

To put the point more formally, from 3-Domain, Binary Independence, Pareto Preference, and Collective P-Transitivity, it follows that there exists an oligarchy, where:

DEFINITION *g is an* oligarchy *iff g satisfies the following two conditions:*

(1) *g is superdecisive.*
(2) *For all* $v \in PROF$ *and* $x, y \in A$, *if* $g \subseteq \{i \mid xP_i^v y \text{ or } yP_i^v x\}$, *and if either* $N - g \subseteq \{i \mid xP_i^v y \text{ or } yP_i^v x\}$ *or else every three-element subset of A is completely free, then* $x\mathrm{P}^v y$ *only if* $xP_g^v y$.

There is no need to assume Nondictatorship—which says, in effect, that no oligarchy has just one member.

THEOREM 3.3.1. *Assume 3-Domain, Binary Independence, Pareto Preference, and Collective P-Transitivity. Then there exists an oligarchy.*

Proof. By 3-Domain and Pareto Preference, *N* is decisive for some pair, whence some subset *g* of *N* is *minimally decisive* for some pair, say *x* vs. *y*, in the sense that *g* is decisive for *x* vs. *y* and no smaller set is decisive for any pair. I shall prove that *g* is an oligarchy. By Lemma 3.2.2, *g* is superdecisive. To show that *g* has property (2) of an oligarchy, let $v \in PROF$ and $x, y \in A$, and suppose:

$$g \subseteq \{i \mid xP_i^v y \quad \text{or} \quad yP_i^v x\},$$

either $N - g \subseteq \{i \mid xP_i^v y \text{ or } yP_i^v x\}$ or else every three-element subset of *A* is completely free,

and

$$x\mathrm{P}^v y;$$

to deduce that $xP_g^v y$. Let

$$g^x = \{i \in g \mid xP_i^v y\} \qquad \text{and} \qquad g^y = \{i \in g \mid yP_i^v x\}.$$

And by 3-Domain, let $z \in A - \{x, y\}$. If $\{x, y, z\}$ is completely free, there is a $w \in PROF$ whose component relations weakly order $\{x, y, z\}$ as follows:

g		
g^x	g^y	$N-g$
x	y	z
y	z	$x \quad y$ ← weakly ordered as in v
z	x	

Otherwise, $N - g \subseteq \{i \mid xP_i^v y \text{ or } yP_i^v x\}$, so the weak orderings just displayed are linear, and thus such a w exists by virtue of 3-Domain. Since $w/\{x, y\} = v/\{x, y\}$ and $x\mathrm{P}^v y$, we have $x\mathrm{P}^w y$ by Binary Independence. But $y\mathrm{P}^w z$ by the superdecisiveness of g. Therefore, $x\mathrm{P}^w z$ by Collective P-Transitivity, and thus g^x is decisive for x vs. z by Theorem 2.3.1. Because g is *minimally* decisive, this would be impossible if g^x were smaller than g. So $g^x = g$. Hence, $xP_g^v y$. ■

Although not itself an impossibility theorem, Theorem 3.3.1 implies one. For it implies that the conditions listed in its hypothesis are inconsistent with:

NONBLOCKER *There is no i such that $\{i\}$ is nearly decisive for every $x \in A$ vs. every $y \in A - \{x\}$.*

In other words, whereas the Arrow conditions with Collective P-Transitivity in place of Collective P + I-Transitivity are consistent, they become inconsistent when we replace Nondictatorship by the stronger but still reasonable Nonblocker (in which case Axiom $\mathbb{P}$ is no longer essential).

THEOREM 3.3.2. *The following five conditions are jointly inconsistent:*
3-Domain, Binary Independence,
Pareto Preference, Nonblocker,
and Collective P-Transitivity.

Proof. Assume the five conditions; to deduce a contradiction. By Theorem 3.3.1, there is an oligarchy, g. By Pareto Preference, $g \neq \emptyset$; say $i \in g$. It suffices to prove that $\{i\}$ is nearly decisive for every $x \in$ A vs. every $y \in$ A $- \{x\}$, contrary to Nonblocker. Since g is an oligarchy, this is trivial if $g = \{i\}$. So suppose $g \neq \{i\}$. By 3-Domain, there is a $v \in$ *PROF* whose component relations linearly order $\{x, y\}$ as follows:

$xP_i^v y$ and $yP_{N-\{i\}}^v x$.

Because $i \in g$, not $yP_g^v x$. Thus, since g is an oligarchy, not $y\mathbb{P}^v x$. Hence, by Theorem 2.3.1, $\{i\}$ is nearly decisive for x vs. y. ■

3.4. TRANSITIVITY CONDITIONS BETWEEN COLLECTIVE P-TRANSITIVITY AND COLLECTIVE P + I-TRANSITIVITY

There are transitivity conditions stronger than Collective P-Transitivity but weaker than Collective P + I-Transitivity. Two such are Collective PIP-Transitivity (the interval-order condition on collective preference) and Collective PIP + IPP-Transitivity (the semiorder condition).

Any transitivity condition involving collective preference and indifference that is stronger than Collective P-Transitivity may be regarded as a conjunction of Collective P-Transitivity with one or more conditions of the following form:

(T) If $x_1\rho_1^v x_2\rho_2^v \cdots \rho_{k-1}^v x_k$ then $x_1\rho_k^v x_k$ (for all $x_1, \ldots, x_k \in A$)

where: $k \geq 3$,

$\rho_i^v \in \{\mathbb{P}^v, \mathbb{I}^v\}$, $i = 1, 2, \ldots, k$,

$\rho_k^v = \mathbb{I}^v$ iff $\rho_i^v = \mathbb{I}^v$ for every $i < k$,

and $\rho_i^v = \mathbb{I}^v$ for some $i < k$.

I have excluded the case in which *every* ρ_i^v is $\mathbb{P}^v$ because (T) would then be a consequence of Collective P-Transitivity.

We can generalize Arrow's Theorem as follows: If, in Arrow's set of conditions, we replace Collective P + I-Transitivity by Collective P-Transitivity plus any condition of the form (T) (one will do), and if we further replace 3-Domain by m-Domain for a suitably large m, we still get an inconsistent set of conditions. By a suitably large m, I mean max(3, $p + 2$), where p is the number of occurrences of "$\mathbb{P}^v$" in the hypothesis of (T). This number is never greater than k, and it is just 3 when (T)'s hypothesis has only one occurrence of $\mathbb{P}^v$ or none.

LEMMA 3.4.1. *Assuming Axiom* $\mathbb{P}$, *any condition* Γ *of the form* (T) *implies a condition of one of the following two forms:*

(1) *If* $x_1\mathbb{I}^v x_2\mathbb{I}^v x_3$ *then* $x_1\mathbb{I}^v x_3$ (for all $x_1, x_2, x_3 \in A$)

(2) *If* $x_1\mathrm{P}^v x_2\mathrm{P}^v \cdots x_{r-1}\ \mathrm{P}^v x_r \mathrm{I}^v y_1 \mathrm{P}^v \cdots y_{t-1}\mathrm{P}^v y_t$ *then* $x_1\mathrm{P}^v y_t$ (for all $x_1, x_2, \ldots, x_r y_1, \ldots, y_t \in A$)
where: $r \geq 1, t \geq 1, r + t \geq 3,$
and $r + t = 2 + |\{i < k \mid \rho_i^v = \mathrm{P}^v\}|.$

Proof. By induction on k. If $k = 3$ then Γ itself has the form (1) or (2), and if only one of the ρ_i^v $(i < k)$ is I^v then Γ itself has the form (2). So suppose $k > 3$ and more than one ρ_i^v $(i < k)$ is I^v. Then Γ has this form:

If $x_1\rho_1^v x_2\rho_2^v \cdots x_{i-1}\rho_{i-1}^v x_i\ \mathrm{I}^v x_{i+1}\rho_{i+1}^v x_{i+2} \cdots \rho_{k-1}^v x_k$ then $x_1\rho_k^v x_k$ (for all $x_1, \ldots, x_k \in A$),

which is equivalent to:

If $x_i\mathrm{I}^v x_{i+1}$ and $x_1\rho_1^v x_2\rho_2^v \cdots x_{i-1}\rho_{i-1}^v x_i$ and $x_{i+1}\rho_{i+1}^v x_{i+2} \cdots \rho_{k-1}^v x_k$ then $x_1\rho_k^v x_k$ (for all $x_1, \ldots, x_k \in A$),

of which the following is a special case (that in which $x_i = x_{i+1}$) and, therefore, a consequence:

If $x_i\mathrm{I}^v x_i$ and $x_1\rho_1^v x_2\rho_2^v \cdots x_{i-1}\rho_{i-1}^v x_i$ and $x_i\rho_{i+1}^v x_{i+2} \cdots \rho_{k-1}^v x_k$ then $x_1\rho_k^v x_k$ (for all $x_1, \ldots, x_k \in A$).

But because $x_i\mathrm{I}^v x_i$ for all $x_i \in A$, this is equivalent to:

If $x_1\rho_1^v x_2\rho_2^v \cdots x_{i-1}\rho_{i-1}^v x_i$ and $x_i\rho_{i+1}^v x_{i+2} \cdots \rho_{k-1}^v x_k$ then $x_1\rho_k^v x_k$ (for all $x_1, \ldots, x_i, x_{i+2}, \ldots, x_k \in A$).

And by inductive hypothesis, this in turn implies some condition of the form (1) or (2). ■

THEOREM 3.4.1. *Let* Γ *be any condition of the form* (T). *Let* $p = |\{i < k \mid \rho_i^v = \mathrm{P}^v\}|$. *The following seven conditions are inconsistent:*
Axiom P, max(3, p + 2)-*Domain,* *Binary Independence,*
Pareto Preference, *Nondictatorship,*
Collective P-Transitivity, *and* Γ.

Proof. Assume the seven conditions; to deduce a contradiction. By Theorem 3.2.1, Collective P + I-Transitivity does not hold. But Collective P-Transitivity holds. So this condition does not hold:

If $x_1\mathrm{I}^v x_2\mathrm{I}^v x_3$ then $x_1\mathrm{I}^v x_3$.

Thus, by Lemma 3.4.1, the following condition holds for some r, t such that $r \geq 1$, $t \geq 1$, and $r + t = \max(3, p + 2)$:

(*) If $x_1 P^v x_2 P^v \cdots x_{r-1} P^v x_r I^v y_1 P^v y_2 P^v \cdots y_{t-1} P^v y_t$ then $x_1 P^v y_t$.

But by hypothesis and Theorem 3.3.1, there is an oligarchy, g. By Pareto Preference, $g \neq \emptyset$; and by Nondictatorship, $|g| \neq 1$. So we may partition g into two nonempty subsets, g_1 and g_2. By max(3, p + 2)-Domain, there are $\hat{x}_1, \ldots, \hat{x}_r, \hat{y}_1, \ldots, \hat{y}_k \in A$ and a $v \in PROF$ of this form:

g		
g_1	g_2	$N - g$
$\hat{x}_1$	$\hat{y}_1$	$\hat{y}_1$
$\vdots$	$\vdots$	$\vdots$
$\hat{x}_r$	$\hat{y}_t$	$\hat{y}_t$
$\hat{y}_1$	$\hat{x}_1$	$\hat{x}_1$
$\vdots$	$\vdots$	$\vdots$
$\hat{y}_t$	$\hat{x}_r$	$\hat{x}_r$

By Pareto Preference,

$\hat{x}_1 P^v x_2 P^v \cdots \hat{x}_{r-1} P^v x_r$.

But because g is an oligarchy and neither $\hat{x}_r P^v_g \hat{y}_1$ nor $\hat{y}_1 P^v_g \hat{x}_r$, we have neither $\hat{x}_r P^v \hat{y}_1$ nor $\hat{y}_1 P^v \hat{x}_r$. Thus,

$\hat{x}_r I^v \hat{y}_1$.

But $\hat{y}_1 P^v \hat{y}_2 P^v \cdots \hat{y}_{t-1} P^v \hat{y}_t$ by Pareto Preference.

Hence $\hat{x}_1 P^v \hat{y}_t$ by (*).

Thus, since g is an oligarchy,

$\hat{x}_1 P^v_g \hat{y}_t$,

contrary to our construction of v. ■

3.5. IMPOSSIBILITY THEOREMS BASED ON COLLECTIVE P-ACYCLICITY

First Theorem: Minimum Resoluteness

Assuming Axiom ℙ, Collective P-Acyclicity is weaker even than Collective P-Transitivity. So if in Arrow's set of conditions we replace Collective P + I-Transitivity by Collective P-Acyclicity, we get a consistent set of conditions. But the inconsistency recurs if we further replace 3-Domain by max(3, n)-Domain, and Nondictatorship by:

NONDESPOTISM *No $i \in N$ is such that $\{i\}$ is decisive for every element of A vs. every other element,*

and add:

MINIMUM RESOLUTENESS *If $yP^v_{N-\{i\}}x$ then either $x\mathrm{P}^v y$ or $y\mathrm{P}^v x$.*

Where Nondictatorship says no one has the power to dictate every pairwise collective choice in all possible situations represented by *PROF*, the slightly stronger Nondespotism says no one has the power to dictate every pairwise collective choice in all such situations *in which he is unanimously opposed.* Minimum Resoluteness says that if at most a single individual prefers x to y while everyone else prefers y to x then there is a definite collective preference one way or the other between x and y—*perhaps* a preference for x, but at any rate *no deadlock*. The idea is that if there is no *other* justification for choosing x in preference to y or for choosing y in preference to x, then the fact that virtually everyone prefers y is reason enough to choose y in preference to x, hence reason enough not merely to toss a coin.

LEMMA 3.5.1. *Assume 3-Domain, Binary Independence, Pareto Preference, Minimum Resoluteness, and Collective P-Acyclicity. Let $i \in N$, and suppose $\{i\}$ is decisive for x vs. y. Then:*
(a) *$\{i\}$ is decisive for x vs. every $z \in A - \{x\}$.*
(b) *$\{i\}$ is decisive for every $z \in A - \{y\}$ vs. y.*
(c) *$\{i\}$ is decisive for every $z \in A$ vs. every $w \in \mathrm{C}A - \{z\}$.*

Proof. (a) Trivial if $z = y$. Otherwise, by 3-Domain, there is a $v \in PROF$ of this form:

$\{i\}$	$n - \{i\}$
x	y
y	z
z	x

xP^vy by the decisiveness of $\{i\}$, and yP^vz by Pareto Preference, whence not zP^vx by Collective P-Acyclicity, so xP^vz by Minimum Resoluteness. Hence, $\{i\}$ is decisive for x vs. z by Theorem 2.3.1.

(b) Trivial if $z = x$. Otherwise, by 3-Domain, some $v \in PROF$ has this form:

$\{i\}$	$n - \{i\}$
z	y
x	z
y	x

zP^vx by Pareto Preference, and xP^vy by the decisiveness of $\{i\}$, whence not yP^vz by Collective P-Acyclicity, so zP^vy by Minimum Resoluteness. Hence, $\{i\}$ is decisive for z vs. y by Theorem 2.3.1.

(c) By 3-Domain, let $t \in A - \{x, z\}$. Then by hypothesis and (a),

	$\{i\}$ is decisive for x vs. t,	
whence	$\{i\}$ is decisive for z vs. t	by (b),
so	$\{i\}$ is decisive for z vs. w	by (a). ■

THEOREM 3.5.1. *The following six conditions are inconsistent:*

max(3, n)-*Domain*, *Binary Independence*,
Pareto Preference, *Nondespotism*,
Minimum Resoluteness, *and Collective P-Acyclicity.*

Proof. Assume the six conditions; to deduce a contradiction. It suffices to deduce Virtual Unanimity, which, by Theorem 3.1.2, is inconsistent with n-Domain (and therewith max(3, n)-Domain) plus Collective P-Acyclicity. So suppose

$$yP^v_{N-\{i\}}x;$$

to prove that yP^vx. If not, then xP^vy by Minimum Resoluteness, whence $\{i\}$ is decisive for x vs. y by Theorem 2.3.1, so $\{i\}$ is decisive for every element of A vs. every other element by Lemma 3.5.1, contrary to Nondespotism. ■

Second Theorem: Positive Responsiveness

The condition max(3, n)-Domain, which says among other things that A has n or more members, is much stronger than the kindred conditions used in previous theorems. We can replace max(3, n)-Domain by 3-Domain in Theorem 3.5.1 and still get an impossibility theorem, however, if we further replace Nondespotism by Nonblocker and Minimum Resoluteness by:

POSITIVE RESPONSIVENESS *If $g \subset h \subseteq N$ and g is nearly decisive for x vs. y then h is decisive for x vs. y,*

and add:

$n \geq 8$.

According to Positive Responsiveness, if a group g is *nearly* decisive for x vs. y, then any more inclusive group is *decisive* for x vs. y. This implies that if a group can dictate the collective indifference of x and y (despite the unanimous opposition of nonmembers), then adding even one member yields a group that can dictate a collective preference for x and over y, so that a single individual can break any tie. Like Minimum Resoluteness, then, Positive Responsiveness is an *anti-tie* condition, although one which specifies a *tie-breaking change* in individuals' preferences rather than a *tie-avoiding combination* of preferences.

The suggested modifications allow us not only to weaken max(3,n)-Domain to 3-Domain but also to weaken Collective P-Acyclicity to Collective P-A*tri*cyclicity, which merely proscribes *three-element* collective-preference cycles.

LEMMA 3.5.1. *Assume 3-Domain, Binary Independence, Pareto Preference, Positive Responsiveness, and Collective P-Atricyclicity. Suppose g is nearly decisive for x vs. y. Then:*
(a) *If $i \in N - g$, $g \cup \{i\}$ is nearly decisive for x vs. any $z \in A - \{x\}$.*
(b) *If $i \in N - g$, $g \cup \{i\}$ is nearly decisive for any $z \in A - \{y\}$ vs. y.*
(c) *If h is a ≥ 4-member subset of $N - g$, $g \cup h$ is decisive for every $z \in A$ vs. every $w \in A - \{z\}$.*

Proof. (a) $g \cup \{i\}$ is decisive for x vs. y by Positive Responsiveness. So (a) is trivial if $z = y$. Otherwise, by 3-Domain, there is a $v \in PROF$ of this form:

$g \cup \{i\}$	$N - (g \cup \{i\})$
x	y
y	z
z	x

xP^vy by the decisiveness of $g \cup \{i\}$, and yP^vz by Pareto Preference, whence not zP^vx by Collective P-Atricyclicity, so $g \cup \{i\}$ is nearly decisive for x vs. z by Theorem 2.3.1.

(b) $g \cup \{i\}$ is decisive for x vs. y by Positive Responsiveness. So (b) is trivial if $z = x$. Otherwise, by 3-Domain, there is a $v \in PROF$ of this form:

$g \cup \{i\}$	$N - (g \cup \{i\})$
z	y
x	z
y	x

zP^vx by Pareto Preference, and xP^vy by the decisiveness of $g \cup \{i\}$, whence not yP^vz by Collective P-Atricyclicity, and thus $g \cup \{i\}$ is nearly decisive for z vs. y by Theorem 2.3.1.

(c) By 3-Domain, let $t \in N - \{x, z\}$. And by hypothesis, let i, j, k be three distinct elements of h. Then by hypothesis and (a),

	$g \cup \{i\}$ is nearly decisive for x vs. t	
whence	$g \cup \{i\} \cup \{j\}$ is nearly decisive for z vs. t	by (b)
so that	$g \cup \{i\} \cup \{j\} \cup \{k\}$ is nearly decisive for z vs. w	by (a)
and thus	$g \cup h$ is decisive for z vs. w by Positive Responsiveness. ■	

LEMMA 3.5.3. *Assume 3-Domain, Binary Independence, Positive Responsiveness, and Collective P-Atricyclicity. Then if g is decisive for some pair and* $|g| \geq 2$, *some two-element subset of g is nearly decisive for some pair.*

Proof. Because g is decisive for some pair, some subset h of g must be *minimally decisive* in this sense: h is decisive for some pair, say x vs. y, and no smaller set is decisive for any pair. If $|h| < 2$, any two-element subset of g that includes h is decisive (hence nearly decisive) for x vs. y by Positive Responsiveness. Otherwise, let i, j be distinct elements of h. By 3-Domain, there is a $z \in A - \{x, y\}$ and a $v \in PROF$ of this form:

$\overbrace{\{i, j\} \quad h - \{i, j\}}^{h}$		$N - h$
z	x	y
x	y	z
y	z	x

xP^vy by the decisiveness of h. So if yP^vz then not zP^vx by Collective P-Atricyclicity, whence $h - \{i, j\}$ is nearly decisive for x vs. z by Theorem 2.3.1, and thus, by Positive Responsiveness, $h - \{i\}$ is decisive for x vs. z, contrary to the minimum decisiveness of h. Consequently, not yP^vz. Hence, by Theorem 2.3.1, $\{i, j\}$ is nearly decisive for z vs. y. ■

LEMMA 3.5.4. *Assume that $n \geq 8$, and assume 3-Domain, Binary Independence, Pareto Preference, Nonblocker, Positive Responsiveness, and Collective P-Atricyclicity. Let i, j, k be three distinct elements of N, and suppose $\{i, j\}$ is nearly decisive for x vs. y. Then $\{j, k\}$ is nearly decisive for some pair.*

Proof. By hypothesis and Positive Responsiveness, $\{i, j, k\}$ is decisive for x vs. y. By Nonblocker, there are $z, w \in A$ such that $\{i\}$ is not nearly decisive for z vs. w. So for some $v \in PROF$, zP_i^vw and $wP_{N-\{i\}}^vz$ and wP^vz. Therefore, by Theorem 2.3.1, $N - \{i\}$ is decisive for w vs. z. Hence, by Lemma 3.5.3 and the assumption that $n \geq 8$, some two-element subset of $N - \{i\}$ is nearly decisive for some pair, and thus, by Lemma 3.5.2 and the assumption that $n \geq 8$, $N - \{i\}$ is decisive for every pair of distinct elements of A.

By 3-Domain, there is a $z \in A - \{x, y\}$ and a $v \in PROF$ of this form:

$\{i\}$	$\{j\}$	$\{k\}$	$N - \{i, j, k\}$
z	x	x	y
x	y	y	z
y	z	z	x

Because $\{i, j, k\}$ is decisive for x vs. y, xP^vy. And because $n - \{i\}$ is decisive for every pair of distinct elements of A, yP^vz. By Collective P-Atricyclicity, then, not zP^vx. Hence, by Theorem 2.3.1, $\{j, k\}$ is nearly decisive for x vs. z. ■

THEOREM 3.5.2. *The following six conditions are inconsistent:*

$n \geq 8$,	*3-Domain,*
Binary Independence,	*Nonblocker,*
Positive Responsiveness, and	*Collective P-Atricyclicity.*

Proof. Assume the six conditions; to deduce a contradiction. By Pareto Preference and 3-Domain, N is decisive for some pair of distinct elements of A. So by Lemma 3.5.3, there are distinct $i, j \in N$ such that $\{i, j\}$ is nearly decisive for some pair. Since $n \geq 8$, let k, m be distinct elements of $N - \{i, j\}$. By Lemma 3.5.4, $\{k, i\}$ is nearly decisive for some pair, say x vs. y, and $\{j, m\}$ is nearly decisive for some pair. So by Lemma 3.5.2, $N - \{k, i\}$ is decisive for every pair of distinct elements of A, hence for y vs. x. But by 3-Domain, there is a $v \in PROF$ such that

$$xP^v_{\{k,\, i\}}y \quad \text{and} \quad yP^v_{N-\{k,\, i\}}x.$$

Because $N - \{k, i\}$ is decisive for y vs. x whereas $\{k, i\}$ is nearly decisive for x vs. y, we have both $y\mathrm{P}^v x$ and not $y\mathrm{P}^v x$. ■

3.6. A REALLY GENERAL IMPOSSIBILITY THEOREM

By using 3-Domain rather than max(3, n)-Domain, Theorem 3.5.2 improves upon Theorem 3.5.1 in one respect. But Theorem 3.5.2 uses two other conditions that are significantly more restrictive than any used earlier. One is that $n \geq 8$. The other is Positive Responsiveness, which requires perfect sensitivity to small changes in individual preferences: when everyone has a definite preference one way or the other between two collectively indifferent alternatives, any individual can break the deadlock by reversing his preference. Although a number of simple pairwise voting rules meet this condition, I find it hard to believe that many complex real-world institutions completely fulfill it. The condition certainly is violated by any collective-choice process, such as the "ratings" system used to decide broadcast-advertising prices, that polls a sample rather than an entire population. Although Positive Responsiveness captures an unimpeachable ideal, an ideal is just what is captures—an ideal rather than a minimum standard of good collective-choice behavior. Yet as far as Theorem 3.5.2 is concerned, it might take just one trivial deviation from this ideal—just one case, say, in which a single individual fails to break a deadlock when he reverses his preference—to ensure fulfillment of Collective P-Acyclicity.

Neither the condition that $n \geq 8$ nor Positive Responsiveness was essential to Theorem 3.5.2, however. The inconsistency remains when we replace the former by the condition that $n \geq 5$ and the latter by:

WEAK POSITIVE RESPONSIVENESS *If g is nearly decisive for x vs. y and h is a $\geq f$-member subset of $N - g$ then $g \cup h$ is decisive for x vs. y,*

where:

DEFINITION $f =$ *the greatest integer* $\leq n/5$.

We cannot further weaken "$n \geq 5$" to "$n \geq 4$," for even the relation of majority preference is perforce acyclic when $n = 4$, $|A| = 3$, and Messrs. $1, 2, \ldots, n$ have linear preferences. Where Positive Responsiveness says that a single individual can break any deadlock, Weak Positive Responsiveness just says that a group comprising a *fifth of society* can break any deadlock.

Two further generalizations of Theorem 3.5.2 are possible. First, we can weaken Nonblocker slightly:

WEAK NONBLOCKER *There is no $i \in N$ such that every subset of N to which i belongs is nearly decisive for every element of A vs. every other element.*

Second, in place of Collective P-Acyclicity or even Collective P-Atricyclicity, all we need assume is that, *for some possible cycle size* (finite or infinite), there is no collective-preference cycle *of that size* with these three features: (1) From any given alternative in the cycle, there is a path of at most four steps leading to any other given alternative (so that movement from any point in the cycle to any other point is relatively easy, and the cycle consists of relatively small subcycles). (2) If the cycle contains more than four alternatives, it contains a *Pareto-inefficient* alternative—one to which some other alternative in the cycle is unanimously preferred. (3) If the given cycle size is $|A|$, then the cycle encompasses all of A—which, of course, is automatically true when A is finite, but not when A is infinite. To put the point another way: The other conditions imply the existence, for some $v \in PROF$, of a P^v-cycle of *any given possible size* with features (1)–(3). In particular, they imply the existence of a P^v-cycle encompassing all of A with features (1) and (2). So if we take A to be the feasible set of the moment, then there is no P^v-undominated feasible alternative, and indeed there is a single P^v-cycle comprising all the feasible alternatives. It is tempting in such cases to equate the choice set with the entire cyclic feasible set. But to do so in this case would be to license a Pareto-inefficient choice if $|A| > 4$. Whereas Theorems 3.5.1 and 3.5.2 just show, for some $v \in PROF$, that *some* potential feasible set contains no P^v-undominated alternative or small "top cycle," the theorem proved below shows this (among other things) for a *fixed* feasible set (which is not even assumed to be finite).

Let γ be the *size of cycles proscribed* by the suggested condition. This can be any cardinal number, finite or infinite, so long as $3 \leq \gamma \leq |A|$, else γ would not be a possible cycle size. It is necessary to replace 3-Domain by γ-Domain. The result is still a more general set of conditions than that involved in Theorem 3.5.2, for I am not assuming that $\gamma > 3$.

Formally, the weakened acyclicity condition runs as follows:

WEAK COLLECTIVE P-ACYCLICITY $3 \leq \gamma \leq |A|$, *and there is no $v \in PROF$ and no γ-member subset B of A such that:*
(1) *for all distinct $x, y \in B$, either xP^vy or $xP^vz_1P^vy$* or
$xP^vz_1P^vz_2P^vy$ or $xP^vz_1P^vz_2P^vz_3P^vy$ *for some $z_1, z_2, z_3 \in B$*;
(2) *if $\gamma > 4$, xP^v_Ny for some $x, y \in B$*; and
(3) *if $\gamma = |A|$, $B = A$.*

Clause (1) says there is a P^v-path of at most four steps leading from any given member of B to any other given member. This implies that B is a P^v-cycle. Clause (2) says that if the given cycle (B) contains more than four alternatives then it contains a Pareto-inefficient alternative.

LEMMA 3.6.1. *Assume Pareto Preference and Weak Collective P-Acyclicity.* (a) *There is no $v \in PROF$ such that, for some distinct x, y, $z \in A$, some $(\gamma - 3)$-member subset B of $A - \{x, y, z\}$, and some linear ordering λ_B* of B,

$$B \cup \{x, y, z\} = A \text{ if } \gamma = |A|,$$

$$P^v_i/B = \lambda_B \text{ for all } i \in N,$$

and $xP^vyP^vzP^vxP^vtP^vy$ *for all $t \in B$.*

That is, there is no $v \in PROF$ *for which* P^v *has this form:*

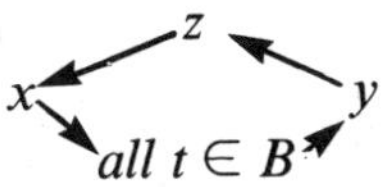

(*where each* P_i^v *linearly orders B the way* λ_B *does*)

(b) *There is no* $v \in PROF$ *such that, for all distinct* $x, y \in A$, *some* $(\gamma - 2)$*-element subset B of* $A - \{x, y\}$, *and some linear ordering* λ_B *of B,*

$$B \cup \{x, y\} = A \text{ if } \gamma = |A|,$$

$$P_i^v/B = \lambda_B \text{ for all } i \in N,$$

and $\quad x\mathrm{P}^v y\mathrm{P}^v t\mathrm{P}^v x$ *for all* $t \in B$.

That is, there is no $v \in PROF$ *for which* P^v *has this form:*

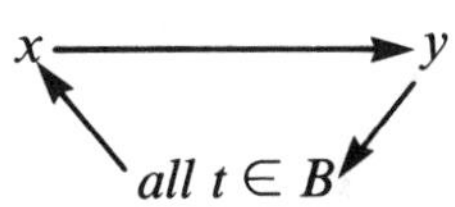

(*where each* P_i^v *linearly orders B the way* λ_B *does*)

Proof. (a) Suppose, on the contrary, that there existed such v, x, y, z, B, and λ_B. Then (1) $|B \cup \{x, y, z\}| = \gamma$; (2) $\mathrm{P}^v/B = \lambda_B$ by Pareto Preference, so that P^v linearly orders B; thus, (3) for all distinct $w, u \in B \cup \{x, y, z\}$, either $w\mathrm{P}^v u$ or $w\mathrm{P}^v t_1\mathrm{P}^v u$ or $w\mathrm{P}^v t_1\mathrm{P}^v t_2\mathrm{P}^v$ or $w\mathrm{P}^v t_1\mathrm{P}^v t_2\mathrm{P}^v t_3\mathrm{P}^v u$ for some $t_1, t_2, t_3 \in B \cup \{x, y, z\}$; and (4) if $|B \cup \{x, y, z\}| > 4$ then there are distinct $w, u \in B$ such that $wP_N^v u$. But this contradicts Weak Collective P-Acyclicity.

The proof of (b) is similar. ■

LEMMA 3.6.2. *Assume* γ*-Domain, Binary Independence, Pareto Preference, Weak Positive Responsiveness, and Weak Collective P-Acyclicity. Suppose g is nearly decisive for x vs. y, h is a* $\geq f$*-member subset of* $N - g$, *and h is not nearly decisive for anything. Then:*

(a) *g is nearly decisive for x vs. every* $z \in A - \{x\}$.

(b) *g is nearly decisive for every* $z \in A - \{y\}$ *vs. y.*

(c) *g is nearly decisive for every* $z \in A$ *vs. every* $w \in A - \{z\}$.

Proof. $g \cup h$ is decisive for x vs. y by hypothesis and Weak Positive Responsiveness.

Since (by Weak Collective P-Acyclicity) $|A| \geq 3$, there is a $(\gamma - 3)$-member subset B of $A - \{x, y, z\}$ such that $B \cup \{x, y, z\} = A$ if $\gamma = |A|$. Let λ_B be a linear ordering of B.

(a) By γ-Domain, there is a $v \in PROF$ of the following form:

g	h	$N - (g \cup h)$
x	λ_B	y
y	z	λ_B
λ_B	x	z
z	y	x

Then

$P_i^v/B = \lambda_B$ for all $i \in N$.

By the decisiveness of $g \cup h$,

$x\mathrm{P}^v y$.

And by Theorem 2.3.1 and the assumption that h is not nearly decisive for anything,

$y\mathrm{P}^v z$ and $y\mathrm{P}^v t$ for all $t \in B$.

But

$t\mathrm{P}^v z$ for all $t \in B$ by Pareto Preference.

So P^v has this form:

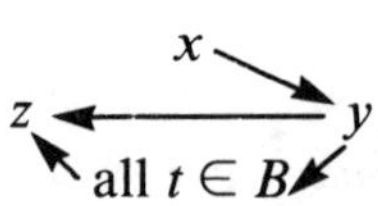

(where each P_i^v linearly orders B the way λ_B does)

By Lemma 3.6.1, then, not $z\mathrm{P}^v x$. Hence, by Theorem 2.3.1, g is nearly decisive for x vs. z.

(b) By γ-Domain, there is a $v \in PROF$ of the following form:

g	h	$N-(g \cup h)$
z	x	y
λ_B	y	z
x	z	λ_B
y	λ_B	x

Then

$$P_i^v/B = \lambda_B \text{ for all } i \in N.$$

By the decisiveness of $g \cup h$,

$$x\mathbb{P}^v y.$$

And by Theorem 2.3.1 and the assumption that h is not nearly decisive for anything,

$$z\mathbb{P}^v x \text{ and } t\mathbb{P}^v x \text{ for all } t \in B$$

But

$$z\mathbb{P}^v t \text{ for all } t \in B \qquad \text{by Pareto Preference.}$$

So $\mathbb{P}^v$ has this form:

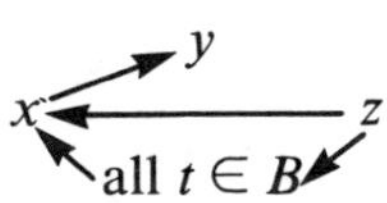

(where each P_i^v linearly orders B the way λ_B does)

By Lemma 3.6.1(a), then, not $y\mathbb{P}^v z$. Hence, by Theorem 2.3.1, g is nearly decisive for z vs. y.

(c) Because $|A| \geq 3$, there is a $t \in A - \{x, z\}$. By hypothesis and (a),

	g is nearly decisive for x vs. t	
whence	g is nearly decisive for z vs. t	by (b)
so	g is nearly decisive for z vs. w	by (a). ■

LEMMA 3.6.3. *Assume that $n \geq 5$, and assume γ-Domain, Binary Independence, Pareto Preference, Weak Positive Responsiveness, and Weak Collective P-Acyclicity. Suppose g is nearly decisive for x vs. y and h is an f-member subset of $N - g$. Then:*

(a) *$g \cup h$ is nearly decisive for x vs. every $z \in A - \{x\}$.*
(b) *$g \cup h$ is nearly decisive for every $z \in A - \{y\}$ vs. y.*
(c) *If $|g| = f$, then for every $z \in A$, either g is nearly decisive for x vs. z or h is nearly decisive for z vs. y.*

Proof. (a) is trivial if $z = y$, (b) if $z = x$, and (c) if $z = x$ or $z = y$. So suppose $z \in A - \{x, y\}$.

$g \cup h$ is decisive for x vs. y by Weak Positive Responsiveness.

Since $|A| \geq 3$, there is a $(\gamma - 3)$-member subset B of $A - \{x, y, z\}$ such that $B \cup \{x, y, z\} = A$ if $g = |A|$. Let λ_B be a linear ordering of B.

(a) By γ-Domain, some $v \in PROF$ has this form:

$g \cup h$	$N - (g \cup h)$
x	y
y	λ_B
λ_B	z
z	x

Then

$P_i^v/B = \lambda_B$ for all $i \in N$.

By the decisiveness of $g \cup h$,

xP^vy.

But by Pareto Preference,

yP^vtP^vz for all $t \in B$,

and

yP^vz.

So P^v *has this form:*

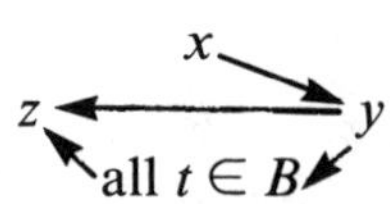

(where each P_i^v linearly orders B the way λ_B does)

By Lemma 3.6.1(a), then, not $z\mathrm{P}^v x$. Hence, by Theorem 2.3.1, g is nearly decisive for x vs z.

(b) By γ-Domain, some $v \in PROF$ has this form:

$g \cup h$	$N - (g \cup h)$
z	y
λ_B	z
x	λ_B
y	x

Then

$P_i^v/B = \lambda_B$ for all $i \in B$.

By the decisiveness of $g \cup h$,

$x\mathrm{P}^v y$.

But by Pareto Preference,

$z\mathrm{P}^v t\mathrm{P}^v x$ for all $t \in B$,

and

$z\mathrm{P}^v x$.

So P^v has this form:

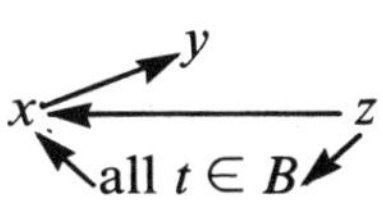

(where each P_i^v linearly orders B the way λ_B does)

By Lemma 3.6.1(a), then, not $y\mathrm{P}^v z$. Hence, by Theorem 3.2.1, g is nearly decisive for z vs. y.

(c) Because $n \geq 5$, $n - (g \cup h)$ has an f-member subset, p, and $N - (g \cup h \cup p)$ has more than f members.

Since g is nearly decisive for x vs. y, $g \cup h$ is nearly decisive for x vs. every $t \in B$ by (a), and for every $t \in B$ vs. y by (b). Thus, by Weak Positive Responsiveness,

$g \cup h$ is decisive for x vs. y,

$g \cup h \cup p$ is decisive for x vs. every $t \in B$,

and

$g \cup h \cup [N - (g \cup h \cup p)]$ is decisive for every $t \in B$ vs. y.

By γ-Domain, some $v \in PROF$ has this form:

g	h	p	$N - (g \cup h \cup p)$
x	z	y	λ_B
λ_B	x	z	y
y	λ_B	x	z
z	y	λ_B	x

Then

$P_i^v/B = \lambda_B$ for all $i \in N$.

But by the decisiveness of $g \cup h$, $g \cup h \cup p$, and $g \cup h \cup [N - (g \cup h \cup p)]$, we have:

$x\mathrm{P}^v y$, and $x\mathrm{P}^v t\mathrm{P}^v y$ for all $t \in B$.

So P^v has the following form:

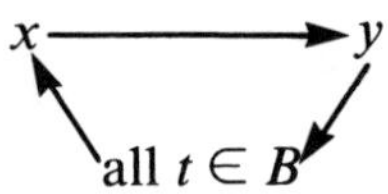

(where each P_i^v linearly orders B the way λ_B does)

Hence, by Lemma 3.6.1(a),

not both $y\mathrm{P}^v z$ and $z\mathrm{P}^v x$.

Thus, by Theorem 2.3.1, either h is nearly decisive for z vs. y or g is nearly decisive for x vs. z. ■

LEMMA 3.6.4. *Assume that* $n \geq 5$, *and assume γ-Domain, Binary Independence, Pareto Preference, Weak Positive Responsiveness, and Weak Collective P-Acyclicity. Suppose* $g \subseteq N$, $|g| \geq 4f$, *and some f-member subset of g is nearly decisive for some pair. Then g is decisive for every* $z \in A$ *vs. every* $w \in A - \{z\}$.

Proof. By hypothesis, we may partition g into four subsets, g_1, g_2, g_3, and g_4, such that

$$|g_1| = |g_2| = |g_3| = f \leq |g_4|$$

and g_1 is nearly decisive, say, for x vs. y.

Because $|A| \geq 3$, there is a $t \in A - \{x, z\}$.

Case 1. $t = w$. So $w \neq x$. Then $g_1 \cup g_2$ is nearly decisive for x vs. w by Lemma 3.6.3(a), whence $g_1 \cup g_2 \cup g_3$ is nearly decisive for z vs. w by Lemma 3.6.3(b), and thus $g_1 \cup g_2 \cup g_3 \cup g_4 = g$ is decisive for z vs. w by Weak Positive Responsiveness.

Case 2. $t \neq w$. So $t \notin \{x, z, w\}$. By Lemma 3.6.3(c), either g_1 is nearly decisive for x vs. t, or g_2 is nearly decisive for t vs. y.

Subcase 2a. g_1 is nearly decisive for x vs. t. Then $g_1 \cup g_2$ is nearly decisive for z vs. t by Lemma 3.6.3(b), whence $g_1 \cup g_2 \cup g_3$ is nearly decisive for z vs. w by Lemma 3.6.3(a), and thus $g_1 \cup g_2 \cup g_3 \cup g_4 = g$ is decisive for z vs. w by Weak Positive Responsiveness.

Subcase 2b. g_2 is nearly decisive for t vs. y. Then $g_1 \cup g_2$ is nearly decisive for t vs. w by Lemma 3.6.3(a), whence $g_1 \cup g_2 \cup g_3$ is nearly decisive for z vs. w by Lemma 3.6.3(b), and thus $g_1 \cup g_2 \cup g_3 \cup g_4 = g$ is decisive for z vs. w by Weak Positive Responsiveness. ■

LEMMA 3.6.5. *Assume that $n \geq 5$, and assume γ-Domain, Binary Independence, Pareto Preference, Weak Positive Responsiveness, and Weak Collective P-Acyclicity. Then there do not exist disjoint sets g, h such that each is nearly decisive for some pair and $|g| \leq f = |h|$.*

Proof. Suppose such sets existed; to deduce a contradiction.

Say g is nearly decisive for x vs. y. But $|N - g| \geq 4f$. So by Lemma 3.6.4, since $h \subseteq N - g$ and h is nearly decisive for some pair, $N - g$ is decisive for y vs. x.

But by γ-Domain, there is a $v \in PROF$ for which $xP^v_g y$ and $yP^v_{N-g}x$. Then by the decisiveness of $N - g$, $y\mathrm{P}^v x$, but by the near decisiveness of g, not $y\mathrm{P}^v x$. ■

LEMMA 3.6.6 *Assume that $n \geq 5$, and assume γ-Domain, Binary Independence, Pareto Preference, Weak Positive Responsiveness, and Weak Collective P-Acyclicity. Then for some $i \in N$, $\{i\}$ is nearly decisive for every $z \in A$ vs. every $w \in A - \{z\}$.*

Proof. Because N is finite and (by Pareto Preference) decisive for any given pair of distinct elements of A, some subset M of N is a *minimal decisive set:* M is decisive for some pair, say x vs. y, and no smaller set is decisive for any pair. I shall first show that some $\leq (f+1)$-member subset of M is nearly decisive for some pair.

Trivial if $|M| < f + 1$. Otherwise, let $i \in M$, and let F be an f-member subset of $m - \{i\}$.

Let B be a $(\gamma - 2)$-member subset of $A - \{x, y\}$ such that $B \cup \{x, y\} = A$ if $\gamma = |A|$, and let λ_B be a linear ordering of B.

By γ-Domain, some $v \in PROF$ has this form:

M			
i	F	$M - (\{i\} \cup F)$	$N - M$
λ_B	λ_B	x	y
x	x	y	λ_B
y	y	λ_B	x

By Theorem 2.3.1, for every $t \in A$, if *not* $t\mathrm{P}^v x$ then $M - (\{i\} \cup F)$ is nearly decisive for x vs. t, whence $[M - (\{i\} \cup F)] \cup F = M - \{i\}$ is decisive for x vs. t by Weak Positive Responsiveness, contrary to the minimal decisiveness of M. Hence,

$$t\mathrm{P}^v x \text{ for all } t \in B.$$

But by the decisiveness of M,

$$x\mathrm{P}^v y.$$

And

$$P_i^v/B = \lambda_B \text{ for all } i \in N.$$

Therefore, by Lemma 3.6.1(b), y does not bear P^v to every $t \in B$. For some $t \in B$, then, not $y\mathrm{P}^v t$. So $F \cup \{i\}$, which has $f + 1$ members, is nearly decisive for t vs. y by Theorem 2.3.1.

Now let L be any *minimum nearly decisive set:* L is nearly decisive for some pair, and no smaller set is nearly decisive for any pair. Because, as just proved, some $\leq (f+1)$-member subset of M is nearly decisive for some pair,

$$|L| \leq f + 1.$$

By Pareto Preference, L has at least one member, say i. I shall prove that $\{i\}$ is nearly decisive for every $z \in A$ vs. every $w \in A - \{z\}$.

Because $n \geq 5$, there is an f-member subset g of $N - L$. Because no set smaller than L is nearly decisive for any pair, if g were nearly decisive for some pair, we should have $|L| \leq f = |g|$, contrary to Lemma 3.6.5. Therefore, g is not nearly decisive for any pair, and thus, by Lemma 3.6.2,

L is nearly decisive for every element of A vs. every other element.

If $\{i\}$ is nearly decisive for some pair, then $\{i\}$ is no smaller than L, so $\{i\} = L$, and thus $\{i\}$ is nearly decisive for every element of A vs. every other element. Hence, it suffices to show that $\{i\}$ is nearly decisive for some pair.

Suppose $\{i\}$ were *not* nearly decisive for any pair; to deduce a contradiction.

Because $n \geq 5$, there are disjoint, f-member subsets h_1, h_2 of $N - L$.

By Weak Positive Responsiveness, since L is nearly decisive for every element of A vs. every other element, $L \cup h_1$ is *decisive* for every element of A vs. every other element.

Let x, y be distinct elements of A.

Let B be a $(\gamma - 2)$-member subset of $A - \{x, y\}$ such that $B \cup \{x, y\} = A$ if $\gamma = |A|$, and let λ_B be a linear ordering of B.

By γ-Domain, some $v \in PROF$ has this form:

L			
$\{i\}$	$L - \{i\}$	h_1	$n - (L \cup h_1)$
λ_B	x	x	y
x	y	y	λ_B
y	λ_B	λ_B	x

Then

$P_i^v/B = \lambda_B$ for all $i \in B$.

But by the decisiveness of $L \cup h_1$,

$x\mathrm{P}^v y$.

And by Theorem 2.3.1, since $\{i\}$ is not nearly decisive for any pair,

$y\mathrm{P}^v t$ for all $t \in B$.

Therefore, by Lemma 3.6.1(b), there is a $t \in B$ for which not $t\mathrm{P}^{v}x$. Thus, by Theorem 2.3.1, $(L - \{i\}) \cup h_1$ is nearly decisive for x vs. t. Hence, by Weak Positive Responsiveness,

$(L - \{i\}) \cup h_1 \cup h_2$ is decisive for x vs. t.

Now let C be a $(\gamma - 2)$-member subset of $A - \{x, t\}$ such that $C \cup \{x, t\} = A$ if $\gamma = |A|$, and let λ_C be a linear ordering of C.

Because $n \geq 5$, $|N - (L \cup h_1 \cup h_2)| \geq f$. So by Weak Positive Responsiveness, since L is nearly decisive for every element of A vs. every other element,

$L \cup [N - (L \cup h_1 \cup h_2)]$ is *decisive* for every element of A vs. every other element.

By γ-Domain, some $v' \in PROF$ has this form:

L: $\{i\}$	L: $L - \{i\}$	h_1	h_2	$N - (L \cup h_1 \cup h_2)$
t	λ_C	x	x	t
λ_C	x	t	t	λ_C
x	t	λ_C	λ_C	x

Then

$P_i^{v'}/C = \lambda_C$ for all $i \in N$.

But by the decisiveness of $L \cup [N - (L \cup h_1 \cup h_2)]$,

$u\mathrm{P}^{v'}x$ for all $u \in C$.

And by the decisiveness of $(L - \{i\}) \cup h_1 \cup h_2$,

$x\mathrm{P}^{v'}t$.

Hence, by Lemma 3.6.1(b), there is a $u \in C$ such that

not $t\mathrm{P}^{v'}u$.

Thus, by Theorem 2.3.1, $L - \{i\}$ is nearly decisive for u vs. t, contrary to the minimum near decisiveness of L. ■

THEOREM 3.6.1. *The following seven conditions are inconsistent:*

$n \geq 5$, *γ-Domain,*

Binary Independence,	*Pareto Preference,*
Weak Positive Responsiveness,	*Weak Nonblocker,*
and	*Weak Collective P-Acyclicity.*

Proof. Assume the seven conditions; to deduce that there is an $i \in N$ such that every subset of N to which i belongs is nearly decisive for every $x \in A$ vs. every $y \in A - \{x\}$, contrary to Weak Nonblocker. By Lemma 3.6.6, there is an $i \in N$ such that $\{i\}$ is nearly decisive for every element of A vs. every other element. Suppose $i \in g \subseteq N$; to deduce that g is nearly decisive for x vs. y.

Let B be a $(\gamma - 2)$-member subset of $A - \{x, y\}$ such that $B \cup \{x, y\} = A$ if $\gamma = |A|$, and let λ_B be a linear ordering of B.

Because $n \geq 5$, we may partition $N - \{i\}$ into $\geq f$-member subsets h_1 and h_2. By Weak Positive Responsiveness, $\{i\} \cup h_1$ and $\{i\} \cup h_2$ are each decisive for every element of A vs. every other element.

By γ-Domain, there is a $v \in PROF$ such that

$xP^v_g y$ and $yP^v_{N-g}x$,

$xP^v_i tP^v_i y$ for all $t \in B$,

$xP^v_{h_1}t$ and $tP^v_{h_2}y$ for all $t \in B$,

$P^v_j/B = \lambda_B$ for all $j \in N$.

Then by the decisiveness of $\{i\} \cup h_1$ and $\{i\} \cup h_2$,

$x\mathrm{P}^v t\mathrm{P}^v y$ for all $t \in B$.

Therefore, by Lemma 3.6.1(b), not $y\mathrm{P}^v x$. By Theorem 2.3.1, then, since $xP^v_g y$ and $yP^v_{N-g}x$, g is nearly decisive for x vs. y. ■

3.7. SUMMARY OF IMPOSSIBILITY THEOREMS

Each of the eight impossibility theorems proved in this chapter asserts the inconsistency of a set of conditions that includes, for some specified value of k:

k-DOMAIN $|A| \geq k$, *and every k-element subset of A is linearly free (for all linear orderings* $\rho_1, \ldots, \rho_n$ *of* α, *there is a* $v \in PROF$ *for which* $v/\alpha = (\rho_1, \ldots, \rho_n)$).

along with one of the following *rationality* conditions:

COLLECTIVE P + I-TRANSITIVITY *If* $x\mathbb{P}^v y\mathbb{P}^v z$ *then* $x\mathbb{P}^v z$, *and if* $x\mathbb{I}^v y\mathbb{I}^v z$ *then* $x\mathbb{I}^v z$ ($\mathbb{P}^v$ *is a* [*strict*] *weak ordering of A for all* $v \in$ *PROF*).

COLLECTIVE P-TRANSITIVITY *If* $x\mathbb{P}^v y\mathbb{P}^v z$ *then* $x\mathbb{P}^v z$ ($\mathbb{P}^v$ *is a strict partial order for all* $v \in$ *PROF*).

The conjunction of Collective P-Transitivity with:

(T) If $x_1\rho_1^v x_2\rho_2^v \cdots \rho_{k-1}^v x_k$ then $x_1\rho_k^v x_k$ $(x_1, \ldots, x_k \in A)$
where: $k \geq 3$,

$\rho_i^v \in \{\mathbb{P}^v, \mathbb{I}^v\}, i = 1, 2, \ldots, k,$

$\rho_k^v = \mathbb{I}^v$ iff $\rho_i^v = \mathbb{I}^v$ for every $i < k$,

and $\rho_i^v = \mathbb{I}^v$ for some $i < k$.

COLLECTIVE P-ACYCLICITY *For no* $v \in$ *PROF is there a* $\mathbb{P}^v$*-cycle* ($\mathbb{P}^v$ *is a suborder for all* $v \in$ *PROF*).

COLLECTIVE P-ATRICYCLICITY *Not* $x\mathbb{P}^v y\mathbb{P}^v z\mathbb{P}^v x$ (*for no* $v \in$ *PROF is there a three-member* $\mathbb{P}^v$*-cycle*).

WEAK COLLECTIVE P-ACYCLICITY $3 \leq \gamma \leq |A|$, *and there is no* $v \in$ *PROF and no* γ*-member subset B of A such that:*
(1) *for all distinct* $x, y \in B$, *either* $x\mathbb{P}^v y$ *or* $x\mathbb{P}^v z_1\mathbb{P}^v y$ *or* $x\mathbb{P}^v z_1\mathbb{P}^v z_2\mathbb{P}^v y$ *or* $x\mathbb{P}^v z_1\mathbb{P}^v z_2\mathbb{P}^v z_3\mathbb{P}^v y$ *for some* $z_1, z_1, z_2, z_3 \in B$;
(2) *if* $\gamma > 4$, $xP_N^v y$ *for some* $x, y \in B$; *and*
(3) *if* $\gamma = |A|$, $B = A$.

and one of the following *size* conditions, each postulating that any coalition of a specified size can always get its way:

ABSOLUTE MAJORITY CONDITION *If* $|\{i\,|xP_i^v y\}| > n/2$ *then* $x\mathbb{P}^v y$.

VIRTUAL UNANIMITY *If* $xP_{N-\{k\}}^v y$ *then* $x\mathbb{P}^v y$.

PARETO PREFERENCE *If* $xP_N^v y$ *then* $x\mathbb{P}^v y$.

Some of the theorems also involved minimum constraints on *n*. Except for the first two, all the theorems involved:

BINARY INDEPENDENCE *If* $v,w \in$ *PROF*, $v/\{x, y\} = w/\{x, y\}$, *and* $x\mathbb{P}^v y$, *then* $x\mathbb{P}^w y$.

plus one of the following *antidictatorial* conditions:

NONDICTATORSHIP *There is no i such that* $\{i\}$ *is superdecisive.*

Table 1.

Theorem	*Constraint on n*	*k = Domain, where k =*	*Size Condition*	*Rationality Condition*	*Binary Independence*	*Antidictatorial Condition*	*Antitie Condition*
3.1.1	$n = 3$ or $n > 4$	3	Absolute Majority	Collective P-Acyclicity	no	none	none
3.1.2	none	n	Virtual Unanimity	Collective P-Acyclicity	no	none	none
3.2.1 (Arrow's)	none	3	Pareto Preference	Collective P + I-Transitivity	yes	Nondictatorship	none
3.3.2	none	3	Pareto Preference	Collective P-Transitivity	yes	Nonblocker	none
3.4.1	none	max(3, $p + 2$), p = number of p's in hyp. of (T)	Pareto Preference	Collective P-Transitivity plus (T)	yes	Nondictatorship	none
3.5.1	none	max(3, n)	Pareto Preference	Collective P-Acyclicity	yes	Nondespotism	Minimum Resoluteness
3.5.2	$n \geq 8$	3	Pareto Preference	Collective P-Atricyclicity	yes	Nonblocker	Positive Responsiveness
3.6.1	$n \geq 5$	γ	Pareto Preference	Weak Collective P-Acyclicity	yes	Weak Nonblocker	Weak Positive Responsiveness

NONBLOCKER *There is no i such that* $\{i\}$ *is nearly decisive for every* $x \in A$ *vs. every* $y \in A - \{x\}$.

NONDESPOTISM *No* $i \in N$ *is such that* $\{i\}$ *is decisive for every element of A vs. every other element.*

WEAK NONBLOCKER *There is no* $i \in N$ *such that every subset of N to which i belongs is nearly decisive for every element of A vs. every other element.*

And each of the last three theorems—Arrovian variations based on acyclicity conditions—involved one of these *antitie* conditions:

MINIMUM RESOLUTENESS *If* $xP_i^v y$ *while* $yP_{N-\{i\}}^v x$ *then either* $x\mathbb{P}^v y$ *or* $y\mathbb{P}^v x$.

POSITIVE RESPONSIVENESS *If* $g \subset h \subseteq N$ *and g is nearly decisive for x vs y then h is decisive for x vs. y.*

WEAK POSITIVE RESPONSIVENESS *If g is nearly decisive for x vs. y and h is a* $\geq f$*-member subset of* $N - g$, *then* $g \cup h$ *is decisive for x vs. y* (f = *the greatest integer* $\leq n/5$).

Table 1 shows the sets of conditions said to be inconsistent by each of the eight impossibility theorems proved in this chapter.

CHAPTER FOUR

EXTENT OF THE PROBLEM

4.1. WHICH PREFERENCE PROFILES GIVE RISE TO CYCLES?

The Classical Voting Paradox shows that Simple Majority Rule can give rise to collective-preference cycles, hence to situations in which every feasible alternative is dominated under the collective-preference relation. How widespread is this phenomenon?

As you know, it is not peculiar to Simple Majority Rule: it afflicts all collective-choice processes that fulfill certain Arrow-type sets of conditions—such as 3-Domain, Binary Independence, Pareto Preference, Weak Nonblocker, and Weak Positive Responsiveness. Nor is the phenomenon peculiar to certain potential feasible sets: according to Theorem 3.6.1, the cycle can be constructed so as to exhaust *any given set* of three or more alternatives—and to have certain other distressing features as well.

The phenomenon of cyclic collective preference *is* peculiar to certain interesting types of preference profile. Although a survey of the large corpus of results in this area would take us far afield, I shall try to convey something of their flavor.

Single Peakedness

If a preference relation satisfies P + I-Transitivity (if it is a weak ordering), we can depict it graphically by listing alternatives along the horizontal axis and letting the vertical axis stand for levels of preferredness. So we can depict a preference for x over y over z in any

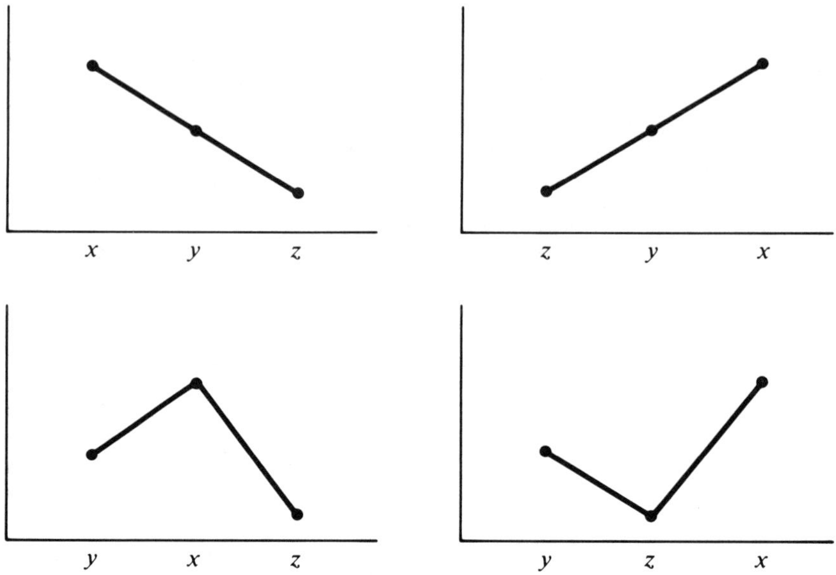

Figure 4.1

of the ways shown in figure 4.1 and others, depending on how we order x, y, and z along the horizontal axis. The first three preference curves are called *single peaked* because they are always up-sloping or always down-sloping or up-sloping to a point and down-sloping thereafter. A preference *profile* is single peaked if there is a way to order the elements of A along the horizontal axis that makes the preference curves of the relations in the profile all single peaked. Here is a precise statement of this condition:

> SINGLE PEAKEDNESS $v \in PROF$, $P^v_1, \ldots, P^v_n$ *weakly order A, and there is a real-valued function f on A such that, for every* $i \in N$ *and for some real number b,* $xP^v_i y$ *for all* $x, y \in A$ *for which either* $b \geq f(x) > f(y)$ *or* $f(y) > f(x) \geq b$.

Think of $f(x)$ as x's position on the horizontal axis and d as Mr. i's "peak." A single-peaked profile represents a one-dimensional or single-issue situation in this sense: Alternatives can be arrayed along a line so that each individual likes alternatives less the farther they are from his ideal point on the line. The line might represent the left-right or liberal-conservative political spectrum, or perhaps quantities of something, such as money, that is to be allocated for some agreed-upon purpose.

Duncan Black, to whom we owe the single-peakedness concept, has proved that if ℙ satisfies SMAC and v satisfies Single Peakedness, then $\mathbb{P}^v$ is transitive, hence acyclic. It follows, of course, that every finite subset α of A has a $\mathbb{P}^v$-undominated element. We can, in fact, say something more definite about $\mathbb{P}^v$-undominated alternatives: Given an ordering of α along the horizontal axis for which Messrs. 1, 2, . . . , n all have single-peaked preference curves, the $\mathbb{P}^v$-undominated elements of α are those elements x such that no majority of individuals have ideal points (peaks) to the left or to the right of x. Such elements lie at or between the *median or medians* (there might be two) of the individuals' ideal points.

More Than One Dimension

Because it specifies a kind of one-dimensional or single-issue situation, Single Peakedness seems highly restrictive. For example, even if voters can agree on how to position candidates for a political office along the liberal-conservative dimension (a big "if"), it would be surprising if that were the only dimension voters cared about.

But when Single Peakedness is generalized even to two dimensions, it no longer guarantees Collective P-Acyclicity. Let alternatives occupy points in two-dimensional space, and suppose each individual likes alternatives less the farther they are from his ideal point in this space. Let $n = 3$, and suppose the profile v represents a situation in which three alternatives, x, y, and z, are positioned relative to the ideal points of the three individuals as in figure 4.2. Because x is closer than y to the ideal points of Messrs. 1 and 3, they prefer x to y. Similarly, Messrs. 2 and 3 prefer y to z, and Messrs. 1 and 2 prefer z to x. So if ℙ satisfies SMAC, $x\mathbb{P}^v y\mathbb{P}^v z\mathbb{P}^v x$, contrary to Collective P-Acyclicity.

To be sure, the feasible set might contain a cycle yet also contain some other alternatives, one of which is unopposed by any majority. If, however, we add to our picture by assuming that every point in two-dimensional space represents an alternative, that the feasible set is

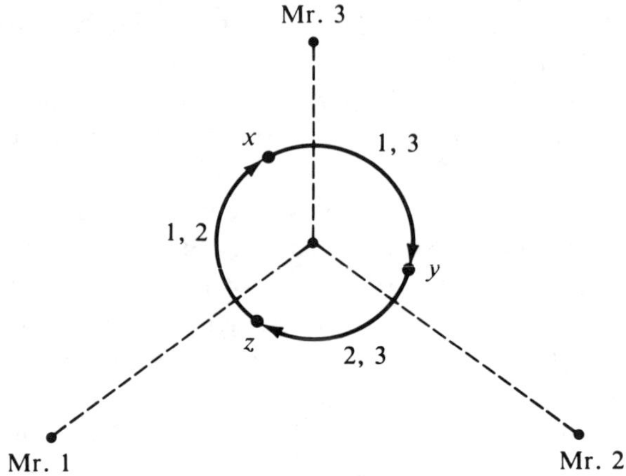

Figure 4.2

either the whole space or a convex subset containing the ideal points of our three individuals, and that no line can be drawn through all three ideal points, then we can deduce that every feasible alternative p is majority dominated: some majority prefer some other feasible alternative to p. For there must exist at least two individuals, say Messrs. 1 and 2, such that p does not lie on the line linking their ideal points. p might, for example, be p_1, p_2, or p_3 in figure 4.3.

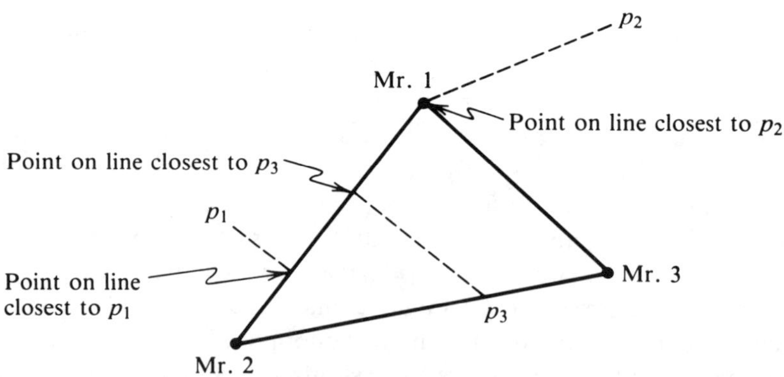

Figure 4.3

Now find the point closest to p on this linking line. Clearly, that point is closer than p to the ideal points of Messrs. 1 and 2, so they prefer that point to p. Thus, p is majority dominated.

Let us generalize this picture by allowing any finite number of individuals. The feasible set is still the whole space or a convex subset containing the individuals' ideal points. To be majority undominated, a feasible alternative must be *a median in all directions*. This means that for any line drawn through it, no majority of individuals have ideal points on one side of that line. To see why, suppose p is a feasible point that is *not* a median in all directions: some majority, g, have their ideal points on one side of some line L through p. Then we can deduce that p is majority dominated. First, draw a line L' parallel to L and between L and g's ideal points, and drop a perpendicular from p to L', as shown in figure 4.4.

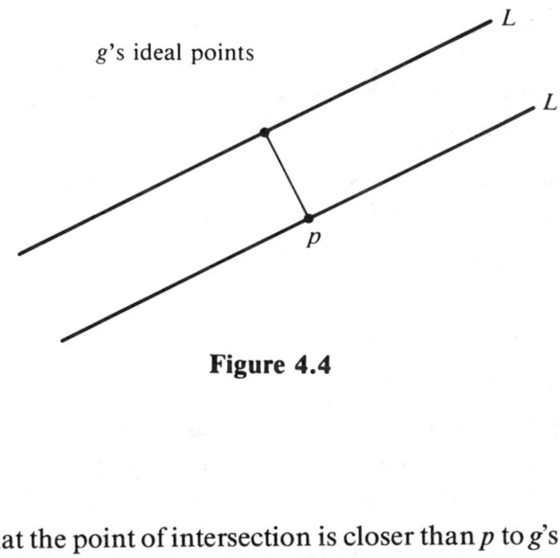

Figure 4.4

Observe that the point of intersection is closer than p to g's ideal points. This means the members of g—who are a majority—all prefer the point of intersection to p. So p is majority dominated.

Since a majority-undominated alternative must be a median in all directions, there need not exist such an alternative. The assumption that there exists a median in all directions is quite strong. For an alternative p to be such a median, there must be no way to divide the population with a line through p so that a majority have ideal points on one side. This condition rules out practically any preferential clustering of individuals in terms of ideology, region, economic interest, political party, or even pure chance.

Nongeometric Generalizations of Single Peakedness

There are other ways to generalize Single Peakedness that preserve Black's Theorem. For one thing, we need only require that Single Peakedness hold for *triples* of alternatives, inasmuch as transitivity in A

Figure 4.5

is equivalent to transitivity in all three-member subsets of A. At the same time, we can allow preference curves of the forms in figure 4.5 to count as single peaked. What we get is the following generalization of Single Peakedness:

> TRIPLE SINGLE PEAKEDNESS $v \in PROF$, α *is a three-member subset of* A, $P^v_1, \ldots, P^v_n$ *weakly order* α, *and there are three distinct* $x, y, z \in \alpha$ *such that, for every* $i \in N$, yP^v_iz *or* yP^v_ix *or* $i \in FLT(v, \alpha)$,

where:

> DEFINITION $FLT(v, \alpha) = \{i \in N \mid v \in PROF, \alpha \in S,$ *and there are no* $x, y \in$ a *for which* $xP^v_iy\}$.

Triple Single Peakedness says there is a way of ordering α (x, y, z) so that each individual (i) either prefers the middle alternative (y) to one or both of the end alternatives (yP^v_iz or yP^v_ix) or has a flat preference curve ($i \in FLT(v, \alpha)$).

This is tantamount to saying that some alternative (y) in α is never bottom ranked except by individuals with flat rankings. Sen and Pattanaik have generalized this to the condition that some alternative in α is never bottom ranked *or* never top ranked *or* never middle ranked except by "flatties":

> VALUE RESTRICTION $v \in PROF$, α *is a three-member subset of* A, $P^v_1, \ldots, P^v_n$ *weakly order* α, *and there are three distinct* $x, y, z \in \alpha$

such that either

(a) *for every* $i \in N - FLT(v, \alpha)$, $yP^v_i z$ *or* $yP^v_i x$; *or*

(b) *for every* $i \in N - FLT(v, \alpha)$, $xP^v_i y$ *or* $zP^v_i y$; *or*

(c) *for every* $i \in N - FLT(v, \alpha)$, *either* $yP^v_i x$ *and* $yP^v_i z$, *or else* $xP^v_i y$ *and* $zP^v_i y$.

Sen and Pattanaik have proved that if $\mathbb{P}$ satisfies SMAC and v and α satisfy Value Restriction then $\mathbb{P}^v$ is transitive in α. And they have proved the same result with Value Restriction replaced by:

LIMITED AGREEMENT $v \in PROF$, α *is a three-member subset of* A, $P^v_1, \ldots, P^v_n$ *weakly order* α, *and there are distinct* $x, y \in \alpha$ *such that* $xP^v_i y$ *for no* $i \in N$.

In both results, Sen and Pattanaik have shown, SMAC can be generalized to Binary Independence plus Neutrality.

Condorcet Freedom: Generalization of Black's Theorem

Here is a condition that generalizes both Value Restriction (hence Single Peakedness) and Limited Agreement:

CONDORCET FREEDOM $v \in PROF$, $\alpha \subseteq A$, $P^v_1, \ldots, P^v_n$ *weakly order* α, *and there do not exist* $i, j, k \in N$ *and three distinct* $x, y, z \in$ a *such that:*

(a) *either* $xP^v_i y$ *and not* $zP^v_i y$, *or* $yP^v_i z$ *and not* $yP^v_i x$;

(b) *either* $yP^v_j z$ *and not* $xP^v_j z$, *or* $zP^v_j x$ *and not* $zP^v_j y$;

(c) *either* $zP^v_k x$ *and not* $yP^v_k x$, *or* $xP^v_k y$ *and not* $xP^v_k z$; *and*

(d) $xP^v_m yP^v_p zP^v_r x$ *for some* $m, p, r \in N$.

If $P^v_1, \ldots, P^v_n$ *linearly* order α, Condorcet Freedom is equivalent to the absence of Condorcet's original three-man, three-alternative example. In general, Condorcet Freedom proscribes profiles of the form:

$$\begin{array}{ccc} i & j & k \\ \hline \left\{\begin{array}{l} x \\ \left.\begin{array}{l} y \\ z \end{array}\right\} \end{array}\right. & \left\{\begin{array}{l} y \\ \left.\begin{array}{l} z \\ x \end{array}\right\} \end{array}\right. & \left\{\begin{array}{l} z \\ \left.\begin{array}{l} x \\ y \end{array}\right\} \end{array}\right. \end{array}$$

where each pair marked by a brace is either a (strict) preference or an indifference, at least one such pair in each ordering is a preference, and each such pair occurs as a preference somewhere in the profile.

This is just Condorcet's example with a certain amount of indifference allowed.

It is evident from inspection of the above diagram that the profiles proscribed by Condorcet Freedom also violate Triple Single Peakedness (and therewith Single Peakedness), Value Restriction, and Limited Agreement. This means that each of the latter conditions implies Condorcet Freedom. But Condorcet Freedom implies none of them. For the following profiles satisfy Condorcet Freedom:

1	2	$N - \{1, 2\}$
x	x	y
y	z	z
z	y	x

1	2	$N - \{1, 2\}$
x	y	zx
y	zx	y
z		

yet the former violates Triple Single Peakedness and Limited Agreement, and the latter violates Value Restriction. So Condorcet Freedom is strictly weaker than each of the other conditions.

Black's Theorem can be generalized as follows: If $\mathbb{P}$ satisfies SMAC and v and α satisfy Condorcet Freedom, $\mathbb{P}^v$ must be transitive in α.

This generalization can be generalized further. SMAC can be replaced by Axiom $\mathbb{P}$ plus Binary Independence and two new conditions. One is

Strong Pareto Preference *If $xP_i^v y$ for some i and $yP_i^v x$ for no i then $x\mathbb{P}^v y$.*

The other new condition I call *Overlap*. To formulate it, I need two definitions, the first of which enables us to talk about a group's power to dictate collective preferences, not only when unanimously opposed by nonmembers, but also when some nonmembers are indifferent:

Definition (a) *p is* pair decisive *for x vs. y iff x, y are distinct elements of A and there are disjoint subsets g, g' of N such that $p = (g, g')$ and, for all $v \in PROF$, if $g = \{i \mid xP_i^v y\}$ and $g' = FLT(v, \{x, y\})$ then $x\mathbb{P}^v y$.* (b) *p is* nearly pair decisive *for x vs. y iff x, y are distinct elements of A and there are disjoint subsets g, g' of N such that $p = (g, g')$ and, for every $v \in PROF$, if $g = \{i \mid xP_i^v y\}$ and $g' =$ FLT$(v, \{x, y\})$ then not $y\mathbb{P}^v x$.*

So an ordered pair of disjoint sets g and g' of individuals is pair decisive for an alternative x vs. another alternative y if x is collectively preferred to y under every profile in which the members of g prefer x to y, the members of g' are indifferent between x and y, and the remaining individuals prefer y to x. And (g, g') is nearly pair decisive for x vs. y if y is not collectively preferred to x under any such profile.

DEFINITION p pair-overlaps q *iff there are sets* g, g', h, h' *such that* $p = (g, g'), q = (h, h')$, *and either* $g \cap (h \cup h') \neq \emptyset$ *or* $h \cap (g \cup g') \neq \emptyset$.

Sets overlap if they share a member. Ordered pairs (g, g') and (h, h') of sets pair-overlap if g shares a member with $h \cup h'$, or h with $g \cup g'$ (it is not enough that g' and h' share a member).

OVERLAP *If* (g, g') *is nearly pair decisive for* x *vs.* y *and* (h, h') *is pair decisive for* z *vs.* x *or for* y *vs.* z, *then* (g, g') *pair-overlaps* (h, h').

This is weaker than the condition—*Strong Overlap*, to give it a name—that if p is nearly pair decisive for something and q is pair decisive for something then p pair-overlaps q. Assuming that any pair (g, g') which is (nearly) pair decisive for something remains so when g or g' is expanded (with disjointness preserved), Strong Overlap is easily seen to be weaker in turn than the condition that (near) pair-decisiveness for one pair of distinct alternatives implies (near) pair decisiveness for all such pairs. And this is weaker still than the Sen-Pattanaik condition of Binary Independence plus Neutrality, which itself is weaker than SMAC.

Unlike Strong Overlap, my Overlap condition is satisfied by the special-majority rules—Two-Thirds, Three-Quarters, and the like—as well as Simple Majority Rule. It is satisfied, indeed, by all pairwise collective-choice processes that meet the

COMPLEMENTARY-QUOTA CONDITION *If* (g, g') *is nearly pair decisive for* x *vs.* y *and* (h, h') *is pair decisive for* z *vs.* x *or for* y *vs.* z *then*

$$\frac{|g|}{|N - g'|} > 1 - \frac{|h|}{|N - h'|}$$

This says that if we take any fraction of nonabstainers that can dictate a collective preference for x over y, or at least block the reverse preference, and if we add this fraction to any fraction that can dictate a collective preference for z over x or for y over z, we get a sum greater than 1. Under Simple Majority Rule, the former fraction must be no less than ½ and the latter greater than ½, so the condition is met. Under

Two-Thirds Majority Rule, it is possible that one of the fractions is just over ⅓. But then the other fraction must be ⅔ or more, so the condition again is met. Similarly for the other special-majority rules.

THEOREM 4.1.1. *Complementary-Quota Condition implies Overlap.*

Proof. Assume Complementary-Quota Condition. Suppose for *reductio* that (g, g') is nearly pair decisive for x vs. y and (h, h') is pair decisive for z vs. x or for y vs. z but (g, g') does *not* pair-overlap (h, h'). Let

$$R = N - (g \cup g' \cup h \cup h').$$

Because (g, g') does not pair-overlap (h, h'), we have

$$N - g' = g \cup h \cup (h' - g') \cup R$$

$$N - h' = h \cup g \cup (g' - h') \cup R.$$

But by Complementary-Quota Condition,

$$\frac{|g|}{|N - g'|} > 1 - \frac{|h|}{|N - h'|}.$$

So

$$\frac{|g|}{|g \cup h \cup (h' - g') \cup R|} > 1 - \frac{|h|}{|h \cup g \cup (g' - h') \cup R|}. \qquad (1)$$

But

$$|g \cup h| \leq |g \cup h \cup (h' - g') \cup R|$$

so

$$\frac{|g|}{|g \cup h|} \geq \frac{|g|}{|g \cup h \cup (h' - g') \cup R|}. \qquad (2)$$

Similarly

$$\frac{|h|}{|g \cup h|} \geq \frac{|h|}{|h \cup g \cup (g' - h') \cup R|}$$

so that

$$1 - \frac{|h|}{|h \cup g \cup (g' - h') \cup R|} \geq 1 - \frac{|h|}{|g \cup h|}$$

and thus

$$\frac{|g|}{|g \cup h|} > 1 - \frac{|h|}{|g \cup h|}. \qquad \text{by (1) and (2).}$$

Hence

$$\frac{|g| + |h|}{|g \cup h|} > 1.$$

But this is possible only if $g \cap h \neq \emptyset$, which contradicts the assumption that (g, g') does not pair-overlap (h, h'). ■

To prove my generalization of Black's Theorem, I need a lemma.

LEMMA 4.1.1. *Suppose x, y are distinct elements of A, $v \in PROF$, $g = \{i \mid xP_i^v y\}$, and $g' =$ FLT$(v, \{x, y\})$. Then* (a) *if $x\mathbb{P}^v y$, (g, g') is pair decisive for x vs. y; and* (b) *if not $y\mathbb{P}^v x$, (g, g') is nearly pair decisive for x vs. y.*

Proof. By hypothesis, g and g' are disjoint subsets of N. Suppose $w \in PROF$, $g = \{i \mid xP_i^w y\}$, and $g' = FLT(w, \{x, y\})$. Then $w/\{x, y\} = v/\{x, y\}$. Thus, by Binary Independence, $x\mathbb{P}^w y$ if $x\mathbb{P}^v y$, which proves (a); and $y\mathbb{P}^v x$ if $y\mathbb{P}^w x$, that is, not $y\mathbb{P}^w x$ if not $y\mathbb{P}^v x$, which proves (b). ■

THEOREM 4.1.2. *Assume Axiom $\mathbb{P}$, Binary Independence, Strong Pareto Preference, and Overlap. Then if v and α satisfy Condorcet Freedom, $\mathbb{P}^v$ is transitive in α.*

Proof. Suppose (as required by Condorcet Freedom) that $v \in PROF$, $\alpha \subseteq A$, and $P_1^v, \ldots, P_n^v$ weakly order α. But suppose $\mathbb{P}^v$ is *not* transitive in α, so that there are $x, y, z \in$ a for which

$$x\mathbb{P}^v y\mathbb{P}^v z \text{ but not } z\mathbb{P}^v z; \qquad (1)$$

I shall deduce that v and α violate Condorcet Freedom. Let

$$g_y^x = \{i \mid xP_i^v y\} \qquad g^{xy} = \text{FLT}(v, \{x, y\})$$

$$g_z^y = \{i \mid yP_i^v z\} \qquad g^{yz} = \text{FLT}(v, \{y, z\})$$

$$g_x^z = \{i \mid zP_i^v x\} \qquad g^{zx} = \text{FLT}(v, \{z, x\}).$$

By (1) and Lemma 4.1.1,

(g_y^x, g^{xy}) is pair decisive for x vs. y,

(g^y_z, g^{yz}) is pair decisive for y vs. z,

and

(g^z_x, g^{zx}) is nearly pair decisive for z vs. x.

So by Overlap, (g^x_y, g^{xy}) pair-overlaps (g^y_z, g^{yz}), which pair-overlaps (g^z_x, g^{zx}), which in turn pair-overlaps (g^x_y, g^{xy}). That is, for some i, j, k,

$$i \in g^x_y \cap (g^z_x \cup g^{yz}) \text{ or } i \in g^y_z \cap (g^x_y \cup g^{xy}),$$

$$j \in g^y_z \cap (g^z_x \cup g^{zx}) \text{ or } j \in g^z_x \cap (g^y_z \cup g^{yz}),$$

and

$$k \in g^z_x \cap (g^x_y \cup g^{xy}) \text{ or } k \in g^x_y \cap (g^z_x \cup g^{zx}).$$

Therefore,

$$\text{either } xP^v_i y \text{ and not } zP^v_i y, \text{ or } yP^v_i z \text{ and not } yP^v_i x; \tag{2}$$

$$\text{either } yP^v_j z \text{ and not } xP^v_j z, \text{ or } zP^v_j x \text{ and not } zP^v_j y; \tag{3}$$

and

$$\text{either } zP^v_k x \text{ and not } yP^v_k x, \text{ or } xP^v_k y \text{ and not } xP^v_k z. \tag{4}$$

Consequently, since P^v_i, P^v_j, and P^v_k weakly order α,

$$xP^v_i z,\ yP^v_j x, \text{ and } zP^v_k y.$$

It follows by Strong Pareto Preference that if $g^z_x = \emptyset$ then $x\mathbb{P}^v z$, contrary to (1). Similarly, if $g^x_y = \emptyset$ or $g^y_z = \emptyset$ then $y\mathbb{P}^v x$ or $z\mathbb{P}^v y$, and so, by Axiom $\mathbb{P}$, not $x\mathbb{P}^v y$ or not $y\mathbb{P}^v z$, contrary to (1). Hence,

$$g^x_y \neq \emptyset,\ g^y_z \neq \emptyset, \text{ and } g^z_x \neq \emptyset;$$

that is, $\qquad xP^v_m yP^v_r zP^v_p x \qquad$ for some m, r, p.

Therefore, by virtue of (2)–(4), v and α violate Condorcet Freedom. ■

In seeking a restriction on preference profiles to block cycles and other failures of transitivity, we need not have looked beyond Condorcet's original three-man, three-alternative example.

Condorcet Effectiveness: A Necessary Condition

Condorcet Freedom is a *sufficient* condition for transitivity and therewith acyclicity of collective preference (given certain assumptions). It is *not* a *necessary* condition: even if Condorcet Freedom is violated, so long as most voters linearly order a triple the same way, the collective preference will order it that way, too. A necessary condition for transitivity or acyclicity must say something, not only about the combination of relations found in a profile, but about their numerical distribution. I shall give an example of such a condition. To avoid distracting complications, I consider only *linear* individual preferences.

If $v \in PROF$ and $x, y, z \in A$, let $M^v(x, y, z)$ be the set of individuals who prefer x to y and y to z in the v situation.

DEFINITION $M^v(x, y, z) = \{i \mid xP_i^v y P_i^v z\}$.

Let v be a preference profile whose component relations are linear in $\{x, y, z\} \subseteq A$, and suppose that $M^v(x, y, z)$, $M^v(y, z, x)$, and $M^v(z, x, y)$ are nonempty. Then v and $\{x, y, z\}$ violate Condorcet Freedom. Even so, $\mathbb{P}^v$ might be transitive in $\{x, y, z\}$. Why? Because the three M sets, although nonempty, might not be large enough to cause trouble. For $\mathbb{P}^v$ to be nontransitive, these sets need not be so large as to be decisive for anything, so long as all twofold unions of them are—so long as $M^v(x, y, z) \cup M^v(y, z, x)$, $M^v(x, y, z) \cup M^v(z, x, y)$, and $M^v(y, z, x) \cup M^v(z, x, y)$ are each decisive for the pair of alternatives for which its members share a preference. Actually, it is enough that two of them be thus decisive and the third nearly so—a condition I call:

CONDORCET EFFECTIVENESS $v \in PROF$, $\alpha \subseteq A$, $P_1^v, \ldots, P_n^v$ *linearly order* α, *and there are* $x, y, z \in \alpha$ *such that* $M^v(x, y, z) \cup M^v(y, z, x)$ *is decisive for* y *vs.* z, $M^v(x, y, z) \cup M^v(z, x, y)$ *is decisive for* x. *vs.* y, *and* $M^v(y, z, x) \cup M^v(z, x, y)$ *is nearly decisive for* z *vs.* x.

Now assume:

MONOTONICITY *If* $g \subseteq h \subseteq N$ *and* g *is decisive or nearly decisive for* x *vs.* y, *so is* h.

Then Condorcet Effectiveness implies that $\mathbb{P}^v$ is not transitive in α. So the *violation* of Condorcet Effectiveness is a *necessary* condition for the transitivity of $\mathbb{P}^v$ in α.

THEOREM 4.1.3. *Assume Monotonicity. Then if* v *and* α *satisfy Condorcet Effectiveness,* $\mathbb{P}^v$ *is not transitive in* α.

Proof. By hypothesis, $P_1^v, \ldots, P_n^v$ linearly order $\alpha \subseteq A$, and there are $x, y, z \in$ a such that

$M^v(x, y, z) \cup M^v(y, z, y)$ is decisive for y vs. z,

$M^v(x, y, z) \cup M^v(z, x, y)$ is decisive for x vs. y,

and

$M^v(y, z, x) \cup M^v(z, x, y)$ is nearly decisive for z vs. x.

Let

$$g_y^x = \{i \mid xP_i^v y\}, \qquad g_z^y = \{i \mid yP_i^v z\}, \qquad \text{and} \quad g_x^z = \{i \mid zP_i^v x\}.$$

Then

$M^v(x, y, z) \cup M^v(z, x, y) \subseteq g_y^x$,

$M^v(x, y, z) \cup M^v(y, z, x) \subseteq g_z^y$,

and

$M^v(y, z, x) \cup M^v(z, x, y) \subseteq g_x^z$.

So by Monotonicity,

g_y^x is decisive for x vs. y,

g_z^y is decisive for y vs. z,

and

g_x^z is nearly decisive for z vs. x.

But since $P_1^v, \ldots, P_n^v$ linearly order $\{x, y, z\} \subseteq \alpha$, we have

$xP_i^v y$ for all $i \in g_y^x$ while $yP_i^v x$ for all $i \in N - g_y^x$,

$yP_i^v z$ for all $i \in g_z^y$ while $zP_i^v y$ for all $i \in N - g_z^y$,

and

$zP_i^v x$ for all $i \in g_x^z$ while $yP_i^v x$ for all $i \in N - g_x^z$.

Consequently, $x\mathrm{P}^v y\mathrm{P}^v z$ but not $x\mathrm{P}^v z$. So P^v is not transitive in α. ■

By themselves, Black's Theorem and Theorems 4.1.2 and 4.1.3 and others of their ilk do not tell us how common collective-preference cycles are in the real world. But they help to show that such phenomena are realistic possibilities—that there is nothing bizarre about the preference profiles that underlie collective-preference cycles.

Consider the 1968 U.S. presidential election. There were three major candidates, Nixon (*N*), Humphrey (*H*), and Wallace (*W*). No doubt almost all who actually voted had linear preference relations on the set $\{N, H, W\}$. Let us regard Messrs. 1, 2, . . . , *n* as those actual voters who did have linear preference relations, and let us assume that any simple majority of such individuals was decisive for every pair of distinct candidates—a harmless simplification of reality.

No doubt many Nixon supporters were ordinary Republicans, disgruntled Democrats, and independents to whom Wallace's regional, racial, law-and-order, and populist appeals were sufficiently unattractive that they preferred Humphrey to Wallace. No doubt many Wallace supporters (including many who decided not to "throw away their vote" on him) were sufficiently conservative or promilitary or pro-law and order or unhappy with the Democratic administration that they preferred Nixon to Humphrey. And no doubt many Humphrey supporters were sufficiently anti-Nixon or were Southern Democrats, lower-middle-class urban workers and ethnics, and others who felt sufficiently alienated or attracted to Wallace's regional, populist, or law-and-order orientation that they preferred Wallace to Nixon. So each of these preferences was well represented within the electorate:

$$\begin{matrix} N & W & H \\ H & N & W \\ W & H & N \end{matrix}$$

Because the electorate's preference profile—call it *v*—contained these preferences, *v* was not Condorcet free. And because the sets $M^v(N, H, W)$, $M^v(W, N, H)$, and $M^v(H, W, N)$ were large, the union of any two of them might have constituted a majority. Although not itself a majority, $M^v(N, H, W)$ could have been large enough to yield a majority when pooled with either of the other sets. That conservative and anti-administration Wallacites ($M^v(W, N, H)$) and anti-Nixon and Wallace-leaning Democrats ($M^v(H, W, N)$) together constituted a majority is less clear but still a very real possibility. So besides not being Condorcet free, the electorate's preference profile could well have been Condorcet effective, giving rise to a collective preference that was nontransitive and undoubtedly cyclic in $\{N, H, W\}$ and therewith a collective choice that was dominated under the collective-preference relation.

In Part IV you will learn that such dominated collective choices are not only realistic possibilities but virtually universal occurrences. Roughly speaking, they occur whenever the final outcome depends essentially on some sort of cooperation or exchange of support or packaging of positions from different issues.

4.2. HOW ESSENTIAL IS BINARY INDEPENDENCE?

Collective-preference cycles are not peculiar to Simple Majority Rule. They are not even peculiar to binary *voting* rules, broadly construed as binary collective-choice processes fulfilling Binary Independence.

For one thing, Theorem 3.2.2, the impossibility theorem based on Virtual Unanimity, did not involve Binary Independence, directly or indirectly. A choice process satisfying Virtual Unanimity need not satisfy Binary Independence. Virtual Unanimity allows pairwise collective choices to reflect individual preferences concerning infeasible alternatives in many cases, prohibiting this only in the extreme case in which all but one individual share a preference.

Weakening Independence

What is more, Theorem 3.5.1 remains true when Binary Independence is drastically weakened, so long as we strengthen max(3, n)-Domain slightly as follows:

DISCRETE max(3, n)-DOMAIN $|A| \geq \max(3, n)$; *and if* α *is a subset of* A *with* max(3, n) *members or fewer and* $\rho_1, \ldots, \rho_n$ *are linear orderings of* α, *there is a* $v \in PROF$ such that:

(a) $v/\alpha = (\rho_1, \ldots, \rho_n)$ *and*
(b) *for all i, x, y, z, if* $xP_i^v yP_i^v z$ and $x, z \in \alpha$, *then* $y \in \alpha$.

This says that every possible combination of n linear orderings of any set α of max(3, n) or fewer alternatives is embodied in some profile in *PROF* that never ranks an alternative foreign to α between two members of α.

Here is the weakened version of Binary Independence:

WEAK BINARY INDEPENDENCE *Suppose*
(a) $v, w \in PROF$ *and* $v/\{x, y\} = w/\{x, y\}$,

(b) $xP^v_k y$ *and* $yP^v_{N-\{k\}}x$, *and*
(c) *there are at most* max(1, n − 2) *objects* z *for which* $xP^v_k zP^v_k y$.
Then if $x\mathbb{P}^v y$, $x\mathbb{P}^w y$.

This is the result of weakening Binary Independence so as to avoid any trace of the objection—unconvincing though it be—raised against that condition in section 2.3. Actually, it is much weaker than need be for this purpose.

Suppose individual preferences between x and y are the same in two situations ($v/\{x, y\} = w/\{x, y\}$), and x is collectively preferred to y in one of the situations ($x\mathbb{P}^v y$). According to Binary Independence, x must be collectively preferred to y in the other situation, too ($x\mathbb{P}^w y$). Weak Binary Independence does not postulate this in general (for *all* $v \in PROF$), nor in the special case in which all but one individual (k) share a preference ($xP^v_k y$ and $yP^v_{N-\{k\}}x$), nor even in the very special case in which the lone dissenter gets his way ($x\mathbb{P}^v y$), but only in that extra-special case in which the lone dissenter, who prefers x to y, gets his way although he ranks x at most max(1, n − 1) notches above y ($xP^v_k zP^v_k y$ for ≤ max (1, n − 2) objects z)—not sufficiently many, when $n > 2$, to outweigh the fact that the n − 1 other individuals all prefer y to x. If $\mathbb{P}$ represents a *reasonable* choice process, one might contend, this extra-special case could never arise anyway. Well and good: then Weak Independence is vacuously satisfied. But if the extra-special case did arise, the collective preference for x over y could not reasonably have resulted from individual preferences concerning infeasible alternatives: the lone dissenter's dictatorial or veto power must have had some other basis—protection of his rights or civil liberties, perhaps. Hence, the collective preference in this case must be independent of individual preferences concerning infeasible alternatives. That, at any rate, is the rationale behind Weak Binary Independence.

LEMMA 4.2.1. *Assume Weak Binary Independence and Minimum Resoluteness. Suppose* $xP^v_k y$, $yP^v_{N-\{k\}}x$, *there are at most* max(1, n − 2) *objects* z *for which* $xP^v_k zP^v_k y$, *and not* $y\mathbb{P}^v x$. *Then* $\{k\}$ *is decisive for* x *vs.* y.

Proof. Suppose $xP^w_k y$ and $yP^w_{N-\{k\}}x$; to prove that $x\mathbb{P}^w y$. Because not $y\mathbb{P}^v x$, Minimum Resoluteness implies that $x\mathbb{P}^v y$, whence, by Weak Binary Independence, $x\mathbb{P}^w y$. ■

LEMMA 4.2.2. *Assume Discrete* max(3, n)*-Domain, Weak Binary Independence, Pareto Preference, Nondespotism, Minimum Resoluteness,* *and Collective P-Acyclicity. Suppose* $xP^v_k y$, $yP^v_{N-\{k\}}x$, *and there are no more than* max(1, $n-2$) *objects z for which* $xP^v_k zP^v_k y$. *Then* $y\mathrm{P}^v x$.

Proof. If *not* $y\mathrm{P}^v x$, then $\{k\}$ would be decisive for x vs. y by Lemma 4.2.1, whence I shall deduce that $\{k\}$ would be decisive for every element of A vs. every other element, contrary to Nondespotism.

For this it suffices to show that if x' and y' are any two elements of A and if $\{k\}$ is decisive for x' vs. y', then (1) $\{k\}$ is decisive for x' vs. every $z \in A - \{x'\}$, (2) $\{k\}$ is decisive for every $w \in A - \{y'\}$ vs. y', and finally (3) $\{k\}$ is decisive for every $t \in A$ vs. every $u \in A - \{t\}$.

Proof of (1). Trivial if $z = y'$. Otherwise, by Discrete max(3, n)-Domain, there is a $v \in PROF$ such that $P^v_1, \ldots, P^v_n$ linearly order $\{x', y', z\}$ as follows:

k	$N-\{k\}$
x'	y'
y'	z
z	x'

and there is no $t \neq y'$ for which $x'P^v_k tP^v_k z$. Because $\{k\}$ is decisive for x' vs. y', $x'\mathrm{P}^v y'$. And by Pareto Preference, $y'\mathrm{P}^v z$. So by Collective P-Acyclicity, not $z\mathrm{P}^v x'$. Hence, by Lemma 4.2.1, $\{k\}$ is decisive for x' vs. z.

Proof of (2). Trivial if $w = x'$. Otherwise, by Discrete max(3, n)-Domain, there is a $v \in PROF$ such that $P^v_1, \ldots, P^v_n$ linearly order $\{x', y', w\}$ as follows:

k	$N-\{k\}$
w	y'
x'	w
y'	x'

and there is no $t \neq x'$ for which $wP^v_k tP^v_k y'$. Because $\{k\}$ is decisive for x' vs. y', $x'\mathrm{P}^v y'$. And by Pareto Preference, $w\mathrm{P}^v x'$. So by Collective P-Acyclicity, not $y'\mathrm{P}^v w$. Hence, by Lemma 4.2.1, $\{k\}$ is decisive for w vs. y'.

Proof of (3). By Discrete max(3, n)-Domain, since $n \geq 3$, there is an $s \in A - \{x', t\}$. By (1), $\{k\}$ is decisive for x' vs. s, whence $\{k\}$ is decisive for t vs. s by (2), so $\{k\}$ is decisive for t vs. u by (1). ■

THEOREM 4.2.1. *The following six conditions are inconsistent:*
Discrete max(3, n)-Domain, *Weak Binary Independence,*
Pareto Preference, *Nondespotism,*
Minimum Resoluteness, *and Collective P-Acyclicity.*

Proof. Assume the six conditions; to deduce a contradiction. By Discrete max(3, n)-Domain, there are $x_1, \ldots, x_n \in A$ and a $v \in PROF$ such that, for every $i \in N$, P_i^v linearly orders $\{x_1, \ldots, x_n\}$ as follows:

$$x_i, x_{i+1}, \ldots, x_n, x_1, x_2, \ldots, x_{i-1},$$

and there is no $y \in A - \{x_1, \ldots, x_n\}$ for which $x_i P_i^v y P_i^v x_{i-1}$.

For each $i \neq 1$ in N, x_n precedes x_1 in the P_i^v-ordering of $\{x_1, \ldots, x_n\}$. That is, $x_n P_i^v x_1$ for each $i \in N - \{1\}$. But $x_1 P_1^v x_n$, and there are just $n - 2$ objects y for which $x_1 P_1^v y P_1^v x_n$. Hence, by Lemma 4.2.2,

$$x_n \mathrm{P}^v x_1.$$

Now let $k \in N - \{n\}$. Then for each $i \in N$, x_k precedes x_{k+1} in the P_i^v-ordering of $\{x_1, \ldots, x_n\}$ unless x_k is the last element, x_{i-1}, in that ordering—unless $i = k + 1$, that is. In short, $x_k P_i^v x_{k+1}$ for each $i \neq k + 1$ in N. But in the P_{k+1}^v-ordering of $\{x_1, \ldots, x_n\}$, x_{k+1} is the first and x_k is the last element. So $x_{k+1} P_{k+1}^v x_k$ and there are just $n - 2$ objects y for which $x_{k+1} P_{k+1}^v y P_{k+1}^v x_k$. Hence, by Lemma 4.2.2, $x_k \mathrm{P}^v x_{k+1}$. Thus, since our choice of k was arbitrary,

$$x_1 \mathrm{P}^v x_2 \mathrm{P}^v \cdots x_{n-1} \mathrm{P}^v x_n \mathrm{P}^v x_1,$$

contrary to Collective P-Acyclicity. ■

Allowing Nontransitive Individual Indifference

Unlike Arrow, I have not assumed that only weak orderings (relations satisfying P + I-Transitivity) can count as individual-preference relations. But neither have I denied this. Let me now deviate from the latter policy slightly by assuming that arbitrary *semiorders* (relations satisfying PIP + IPP-Transitivity but not *necessarily* the slightly stronger P + I-Transitivity) of at least one three-member set can count as individual-preference relations on that set:

MINIMUM SEMIORDER CONDITION *There exists a three-member subset* α *of* A *such that, for all (strict) semiorders* $\rho_1, \ldots, \rho_n$ *of* α, $v/\alpha = (\rho_1, \ldots, \rho_n)$ *for some* $v \in PROF$.

Then following Dennis Packard, we can prove an impossibility theorem that does not involve Binary Independence or any remnant thereof—or much else, for that matter.

THEOREM 4.2.2. *Assuming $n \geq 3$, these three conditions are inconsistent:*
Minimum Semiorder Condition, Strong Pareto Preference, and Collective P-Atricyclicity.

Proof. Assume that $n \geq 3$ and the three conditions hold; to deduce a contradiction. By Minimum Semiorder Condition, since $n \geq 3$, there are $x, y, z \in$ A and a $v \in PROF$ such that

$xP_1^v y$ but neither $xP_1^v z$ nor $zP_1^v y$,

$yP_2^v z$ but neither $yP_2^v x$ nor $xP_2^v z$,

and for all $i \in N - \{1, 2\}$,
$zP_i^v x$ but neither $zP_i^v y$ nor $yP_i^v x$.
Then

$xP_1^v y$ but $yP_i^v x$ for no $i \in N$,

$yP_2^v z$ but $zP_i^v y$ for no $i \in N$,

and

$zP_3^v x$ but $xP_i^v z$ for no $i \in N$.

So by Strong Pareto Preference, $x\mathrm{P}^v y\mathrm{P}^v z\mathrm{P}^v x$, contrary to Collective P-Acyclicity. ■

4.3. WHAT IF COLLECTIVE CHOICES REFLECTED INDIVIDUAL-PREFERENCE INTENSITIES?

Contrary to common dogma, collective-preference cycles are not peculiar to collective-choice processes that are insensitive (or only slightly sensitive) to interpersonal comparisons of preference intensity.

For one thing, none of the assumptions involved in the last two impossibility theorems requires collective choices to be independent of interpersonal comparisons of preference intensity. And as you saw in section 2.2, this is not required, either, by the mere fact that P is a function of preference profiles rather than utility profiles.

What is more, we can represent binary collective choices as a function of utility profiles rather than preference profiles and allow such choices to reflect interpersonal comparisons of preference intensity to a considerable degree and still prove an impossibility theorem similar to Theorem 3.5.2.

Let *UPRF* be the set of *utility profiles*, each an ordered n-tuple $(u_1, \ldots, u_n)$ of real-valued functions on A, representing a possible situation in which (1) u_i is Mr. i's utility measure; (2) $u_i(x) > u_i(y)$ iff Mr. i prefers x to y; and (3) if interpersonal comparisons of preference intensity are possible (something I *allow* but do *not assume*) then $u_i(x) - u_i(y) = r[u_j(y) - u_j(x)] > 0$ and $r > 1$ only if Mr. i prefers x to y more intensely than Mr. j prefers y to x—r *times* more, if *numerically specific* comparisons are possible.

For $v \in UPRF$ and $i \in N$, let U_k^v be the ith utility function in v (just as P_i^v is the ith preference relation in v, for each $v \in PROF$).

Definition $U_i^v = \{(x, y) \mid i \in N$, *and for some* $f_1, \ldots, f_n$, $v = (f_1, \ldots, f_n) \in UPRF$ *and* $f_i(x) = y\}$.

I shall not assume that every ordered n-tuple of real-valued functions on A belongs to *UPRF*, any more than I assumed that every ordered n-tuple of weak orderings of A belongs to *PROF*. Instead I assume there are real numbers a and b fulfilling this condition:

Utility Condition $|A| \geq n \geq 3$, $b > 0$, *and for every n-member subset* α *of* A *and all functions* $f_1, \ldots, f_n$ *on* α *into* $\{a, a + b, a + 2b, \ldots, a + nb\}$, *there is a* $v \in UPRF$ *such that* $U_i^v(x) = f_i(x)$ *for every* $x \in$ a *and* $i \in N$.

Think of b this way: A preference intensity of b units or more is *significant*—big enough to care about.

Let us represent a binary collective-choice process by a function $\mathbb{B}$ that assigns, to each utility profile v, the relation $\mathbb{B}^v$ of collective preference in the v situation.

I shall prove an impossibility theorem involving Utility Condition plus these numerical analogs of some conditions used earlier:

Pareto Preference 2 *If* $U_i^v(x) > U_i^v(y)$ *for all* $i \in N$, *then* $x\mathbb{B}^v y$.

Nondespotism 2 *There is no despot for every element of* A *vs. every other element,* where:

DEFINITION *k is a* despot *for x vs. y iff* $k \in N$, *x and y are distinct elements of A, and for every* $v \in UPRF$, *if* $U_k^v(x) > U_k^v(y)$ *while* $U_i^v(y) > U_i^v(x)$ *for all* $i \in N - \{k\}$ *then* $x\mathrm{B}^v y$.

MINIMUM RESOLUTENESS 2 If $U_k^v(x) > U_k^v(y)$ *while* $U_i^v(y) > U_i^v(x)$ *for all* $i \in N - \{k\}$ *then either* $x\mathrm{B}^v y$ *or* $y\mathrm{B}^v x$.

COLLECTIVE P-ACYCLICITY 2 *There are no* B^v cycles.

MINIMUM INDEPENDENCE *Suppose*

(a) $U_k^v(x) > U_k^v(y)$ *while* $U_i^v(y) - U_i^v(x) \geq b$ *for all* $i \in N - \{k\}$,
(b) $U_k^v(y) \geq U_i^v(x)$ *and* $U_k^v(x) \leq U_j^v(y)$ *for some i, j,*
(c) $\sum_{i=1}^{n} U_i^v(y) \geq \sum_{i=1}^{n} U_i^v(x)$, *and*
(d) $U_k^w(x) > U_k^w(y)$ *while* $U_i^w(y) > U_i^w(x)$ *for all* $i \in N - \{k\}$.

Then if $x\mathrm{B}^v y$, $x\mathrm{B}^w y$.

The first four conditions are straightforward analogs of Pareto Preference, Nondespotism, Collective P-Acyclicity, and Minimum Resoluteness.

In the present context, Binary Independence might be formulated as follows:

Suppose, for every $i \in N$, *that* $U_i^v(x) > U_i^v(y)$ *iff* $U_i^w(x) > U_i^w(y)$, *and* $U_i^v(y) > U_i^v(x)$ *iff* $U_i^w(y) > U_i^w(x)$. *Then if* $x\mathrm{B}^v y$, $x\mathrm{B}^w y$.

This says binary collective choices are independent, not only of individual preferences concerning infeasible alternatives, but also of individual *preference intensities*.

Minimum Independence preserves a remnant of the Binary Independence idea while permitting collective choices to be quite sensitive to interpersonal comparisons of preference intensity, even numerically specific ones. Suppose individual preferences between two alternatives are the same in two situations—although *intensities* of preference may be different—and *x* is collectively preferred to *y* in one of the situations. According to Binary Independence, *x* must be collectively preferred to *y* in the other situation, too. Minimum Independence does not postulate this in general, nor in the special case in which all but one individual share a preference, nor even in the very special case in which the lone dissenter gets his way, but only in that extra-special case in which the dissenter, who prefers x to y, gets his way although (1) everyone else has a significant preference (one with an intensity no less than b) for y over x, (2) the dissenter would be no worse off under y than someone else would be under x and no better off under x than someone else would be under y, and (3) the intensity of his preference is no greater than the aggregate intensity of the others' preferences. In such a case (if indeed there be such a case), the collective preference for x over y does not reflect preference intensities anyway. So requiring the same collective preference in the other situation, in which individual preferences are the same although intensities may differ, does not preclude sensitivity to interpersonal comparisons of preference intensity.

Readers not blessed with your intellect might object to Minimum Resoluteness 2 for this reason: Consider a situation, represented by a utility profile v, in which one individual prefers x to y while the others all prefer y to x with an aggregate intensity exactly as great as his, so that

$$\sum_{i=1}^{n} U_i^v(x) = \sum_{i=1}^{n} U_i^v(y).$$

Why should either of the two alternatives be collectively preferred to the other, as Minimum Resoluteness 2 requires? Why can they not be *collectively indifferent*?

For one thing, it is utterly unrealistic to expect anything but a collective preference for y over x when all but one individual prefer y to x, unless the lone dissenter has been given some special right or autonomy or veto power to decide the x-vs.-y issue, in which case x will be collectively preferred to y—still no collective indifference. In other words, it is unrealistic to expect a mere coin toss when virtually all individuals share a preference.

What is more, in the absence of any *other* reason for making a definite choice one way or the other between x and y, the fact that virtually everyone prefers y to x is reason enough, surely, to decide the issue in y's favor ("The greatest good *for the greatest number*").

Finally, even if interpersonal comparisons of preference intensity were possible, they would doubtless be numerically inspecific, or at least somewhat imprecise. But so long as such comparisons are either unavailable or numerically inspecific or even slightly imprecise, equality of aggregate utility numbers does not guarantee *equality* of preference intensity, but at most only *near* equality, so we have every reason to look at other factors—at how many people prefer which alternative, for example—*even if* equality of preference intensity *were* a sufficient condition for collective indifference. My contention, after all, is that the conditions above allow collective choices to reflect interpersonal comparisons of preference intensity *to a considerable degree*—limited, perhaps, by measurement imprecision.

LEMMA 4.3.1. *Assume Minimum Resoluteness 2 and Minimum Independence, and assume that* $b > 0$ *and* $n \geq 3$. *Suppose*

(a) $U_k^v(x) - U_k^v(y) = nb - b,$

(b) $U_i^v(y) - U_i^v(x) = b$ *for all* $i \in N - \{k\},$

(c) $U_i^v(y) = U_k^v(x)$ *and* $U_j^v(x) = U_k^v(y)$ *for some i and j, and*

(d) *not* $y\mathrm{B}^v x$.

Then k is a despot for x vs. y.

Proof. Suppose

$$U_k^w(x) > U_k^w(y) \text{ while } U_i^w(y) > U_i^w(x) \text{ for all } i \in N - \{k\};$$

to prove that $x\mathrm{B}^w y$. Since $n \geq 3$ and $b > 0$, (a) and (b) imply:

$$U_k^v(x) > U_k^v(y) \text{ while } U_i^v(y) > U_i^v(x) \text{ for all } i \in N - \{k\},$$

whence

$$x\mathrm{B}^v y$$

by (d) and Minimum Resoluteness 2. But

$$\sum_{i \in N}^{n} U_i^v(y) - \sum_{i \in N}^{n} U_i^v(x) = \sum_{i \in N}^{n} [U_i^v(y) - U_i^v(x)]$$

$$= \sum_{i \in N-\{k\}}^{n} [U_i^v(y) - U_i^v(x)] + [U_k^v(y) - U_k^v(x)]$$

$$= \sum_{i \in N-\{k\}}^{n} [U_i^v(y) - U_i^v(x)] - [U_k^v(x) - U_k^v(y)]$$

$$= (n - 1)b - (nb - b) \qquad \text{by (a) and (b)}$$

$$= (nb - b) - (nb - b) = 0,$$

whence

$$\sum_{i\,N}^{n} U_i^v(y) = \sum_{i\,N}^{n} U_i^v(x).$$

So by Minimum Independence, $x\mathrm{B}^w y$. ■

LEMMA 4.3.2. *Assume Utility Condition, Minimum Independence, Pareto Preference 2, Nondespotism 2, Minimum Resoluteness 2, and Collective P-Acyclicity 2. Suppose*

(a) $U_k^v(x) = a + nb$,
(b) $U_k^v(y) = a + b$,
(c) $U_i^v(y) - U_i^v(x) = b$ *for all* $i \in N - \{k\}$,
(d) $U_i^v(y) = a + nb$ *for some i, and*
(e) $U_i^v(x) = a + b$ *for some i.*

Then $y\mathrm{B}^v x$.

Proof. By (a) and (b), $U_k^v(x) - U_k^v(y) = nb - b$. And by (d) and (e), $U_i^v(y) = U_k^v(x)$ and $U_j^v(x) = U_k^v(y)$ for some i, j. So by Lemma 4.3.1, if *not* $y\mathrm{B}^v x$, then k would be a despot for x vs. y, whence I shall deduce that k would be a despot for every element of A vs. every other element, contrary to Nondespotism 2.

For this it suffices to show that if x' and y' are any two elements of A and if k is a despot for x' vs. y', then (1) k is a despot for x' vs. every $z \in A - \{x'\}$, (2) k is a despot for every $w \in A - \{y'\}$ vs. y', and finally (3) k is a despot for every $t \in A$ vs. every $s \in A - \{t\}$.

Proof of (1). Trivial if $z = y'$. Otherwise, by Utility Condition (which implies that $n \geq 3$), there is an $i \in N - \{k\}$ and a $v \in UPRF$ such that

$U_k^v(x') = a + nb - b$ and $U_k^v(z) = a$,

$U_i^v(z) = a + nb - b$ and $U_i^v(x) = a + nb - 2b$,

$U_j^v(x) = a$ and $U_j^v(z) = a + b$ for all $j \in N - \{k, i\}$,

$U_k^v(y) = a + b$,

$U_i^v(y) = a + nb$,

and

$U_j^v(y) = a + 2b$ for all $j \in N - \{i, k\}$.

Then $U_k^v(x') > U_k^v(y')$ while $U_j^v(y') > U_j^v(x')$ for all $j \in N - \{k\}$. Thus, since k is a despot for x' vs. y', $x'\mathrm{B}^v y'$. But $U_j^v(y') > U_j^v(z)$ for all $j \in N$, whence $y'\mathrm{B}^v z$ by Pareto Preference 2. Hence, by Collective P-Acyclicity 2,

not $z\mathrm{B}^v x'$.

But

$U_k^v(x') - U_k^v(z) = nb - b,$

$U_j^v(z) - U_j^v(x') = b$ for all $j \in N - \{k\}$,

$u_i^v(z) = U_k^v(x'),$

and

$U_j^v(x') = U_k^v(z)$ for all (hence for some) $j \in N - \{i, k\}$).

By Lemma 4.3.1, then, k is a despot for x' vs. z.

Proof of (2). Trivial if $w = x'$. Otherwise, by Utility Condition, there is an $i \in N$ and a $v \in UPRF$ such that

$U_k^v(w) = a + nb$ and $U_k^v(y') = a + b,$

$U_i^v(y') = a + nb$ and $U_i^v(w) = a + nb - b,$

$U_j^v(y') = a + 2b$ and $U_j^v(w) = a + b$ for all $j \in N - \{k\}$,

$U_k^v(x') = a + nb - b,$

$U_i^v(x') = a + nb - 2b,$

and

$U_j^v(x') = a$ for all $j \in N - \{k, i\}$.

Then $U_k^v(x') > U_k^v(y')$ while $U_j^v(y') > U_j^v(x')$ for all $j \in N - \{k\}$. Thus, since k is a despot for x' vs. y', $x'\mathrm{B}^v y'$. But $U_j^v(w) > U_j^v(x')$ for all $j \in N$, whence $w\mathrm{B}^v x'$ by Pareto Preference 2. Hence, by Collective P-Acyclicity 2,

not $y'\mathrm{B}^v w$.

But

$$U_k^v(w) - U_k^v(y') = nb - b,$$

$$U_j^v(y') - U_j^v(w) = b \text{ for all } j \in N - \{k\},$$

$$U_i^v(y') = U_k^v(w),$$

and

$$U_j^v(w) = U_k^v(y') \text{ for all (hence for some) } j \in N - \{i, k\}.$$

By Lemma 4.3.1, then, k is a despot for w vs. y'.

Proof of (3). Since (by Utility Condition) $|A| \geq 3$, let $r \in A - \{x', t\}$. By hypothesis and (1), k is a despot for x' vs. r, whence k is a despot for t vs. r by (2), so k is a despot for t vs. s by (1). ■

THEOREM 4.3.1. *The following six conditions are inconsistent:*
Utility Condition, *Minimum Independence,*
Pareto Preference 2, *Nondespotism 2,*
Minimum Resoluteness 2, *and* *Collective P-Acyclicity 2.*

Proof. Assume the seven conditions; to deduce a contradiction. By Utility Condition, there are $x_1, \ldots, x_n \in A$ and a $v \in UPRF$ such that, for all $j, r \in N$,

$$U_j^v(x_r) = \begin{cases} a + jb - rb \text{ if } 1 \leq r < j \leq n, \\ a + jb - rb + nb \text{ if } 1 \leq j \leq r \leq n. \end{cases}$$

Then whenever $1 \leq i \leq n$, we have

$$U_i^v(x_n) = a + ib - nb + nb = a + ib$$

and

$$U_i^v(x_1) = a + ib - b.$$

So

$$U_i^v(x_n) - U_i^v(x_1) = b \text{ for all } i \in N - \{1\}.$$

But

$$U_1^v(x_1) = a + b - b + nb = a + nb,$$

$$U_1^v(x_n) = a + b + nb - nb = a + b,$$

$$U_2^v(x_1) = a + 2b - b = a + b,$$

and

$$U_n^v(x_n) = a + nb - nb + nb = a + nb.$$

Therefore, by Lemma 4.3.1,

$$x_n \mathrm{B}^v x_1. \tag{*}$$

Now let $k \in N - \{i\}$. Then whenever $1 \le i < k \le n$, we have

$$U_i^v(x_{k-1}) = a + ib - (k-1)b + nb$$

and

$$U_i^v(x_k) = a + ib - kb + nb,$$

whence

$$U_i^v(x_{k-1}) - U_i^v(x_k) = b.$$

And whenever $1 < k < i \le$ m, we have

$$U_i^v(x_{k-1}) = a + ib - (k-1)b$$

and

$$U_i^v(x_k) = a + ib - kb,$$

whence

$$U_i^v(x_{k-1}) - U_i^v(x_k) = b.$$

So

$$U_i^v(x_{k-1}) - U_i^v(x_k) = b \text{ for all } i \in N - \{k\}.$$

But

$$U_k^v(x_k) = a + (k-1)b - kb = a + b,$$

$$U_k^v(x_{k-1}) = a + (k-1)b - kb = a + b,$$

$$U_{k-1}^v(x_{k-1}) = a + (k-1)b - (k-1)b + nb = a + nb,$$

and

$$U_{k+1}^v(x_k) = a + (k+1)b - kb = a + b \text{ if } k < n,$$

whereas

$$U_1^v(x_k) = a + b - nb + nb = a + b \text{ if } k = n.$$

By Lemma 4.3.2, then, $x_{k-1}\mathrm{B}^v x_k$. Thus, since our choice of k was arbitrary,

$$x_{k-1}\mathrm{B}^v x_k \text{ for all } k \in N - \{1\}.$$

Hence, by (*),

$$x_1\mathrm{B}^v x_2\mathrm{B}^v \cdots x_{n-1}\mathrm{B}^v x_n\mathrm{B}^v x_1,$$

contrary to Collective P-Acyclicity 2. ■

Because I never assumed that individual utility measures have more than ordinal significance (I just allowed the possibility that they do), Theorem 4.3.1 is really a generalization of Theorem 3.5.1.

4.4. WHAT ABOUT INDIVIDUAL CHOICE?

Cycles are not peculiar to collective as opposed to individual preference. The impossibility theorems I have proved can be interpreted in terms of individual as well as collective pairwise-choice processes.

Let $\mathbb{P}$ represent, not a *collective* choice process, but rather the rule, mechanism, or process by which a *single individual*, Bertha, makes pairwise choices between members of A. For $v \in PROF$, $\mathbb{P}^v$ is Bertha's preference relation in the v-situation—the relation of one alternative to another when the rule or mechanism by which Bertha makes choices prescribes or produces the choice of x over y, given a choice between the two.

$P_1^v, \ldots, P_n^v$ might still be the preference relations of Messrs. 1, 2, . . . , n in the v situation. In that case, Bertha wishes to choose in such a way as to take account of those individuals' preferences. Messrs. 1, 2, . . . , n are Bertha's friends, relatives, employees, customers, countrymen, or whoever.

Or we can interpret $P_1^v, \ldots, P_n^v$ as the relations of better to worse, in the v situation, according to each of n *criteria*, $c_1, \ldots, c_n$, so that xP_i^vy if x is better than y according to criterion c_i. If, for example, A comprises motor cars, $c_1, \ldots, c_n$ include such criteria as speed, safety, comfort, and resale value.

It strikes me as reasonable and realistic to expect ℙ, as just interpreted, to satisfy SMAC *in some cases*—particularly when all Bertha knows about two alternatives is which is the better according to what criteria, and there is no basis, in the circumstances, for differential weighting of criteria of for measuring degrees of criterial preferability, except arbitrarily. We might expect ℙ to satisfy Virtual Unanimity even more often, and to satisfy the Arrovian conditions of Pareto Preference, Nondictatorship, and Binary Independence more often still. Likewise for Nondespotism, Nonblocker, Minimum Resoluteness, and Weak Positive Responsiveness.

Some might question the reasonableness of Binary Independence. But my only contention here is that Binary Independence is a reasonable condition on Bertha's choice process *in some cases*. Besides, all one really needs in order to show that Bertha's preferences can be cyclic is Weak Binary Independence or Minimum Independence.

To illustrate how Bertha's preferences can by cyclic, let $n = 5$, and suppose, in the situation represented by the preference profile v, five alternatives are ranked as follows by the five criteria:

c_1	c_2	c_3	c_4	c_5
x_1	x_2	x_3	x_4	x_5
x_2	x_3	x_4	x_5	x_1
x_3	x_4	x_5	x_1	x_2
x_4	x_5	x_1	x_2	x_3
x_5	x_1	x_2	x_3	x_4

Suppose Bertha treats no criterion as dominant, although she may weight them differentially. And suppose Bertha has no basis for making *intercriterial preferability comparisons*, or *tradeoffs*—judgments like "y's preferability to z according to c_i is [r times] greater than z's preferability to y according to c_j." Then since x_5 is better than x_1 according to all but one criterion, it seems reasonable for Bertha to choose x_5 over x_1, given a choice between the two. So we may suppose $x_5\mathrm{P}^v x_1$. Similarly, $x_1\mathrm{P}^v x_2\mathrm{P}^v x_3\mathrm{P}^v x_4\mathrm{P}^v x_5$, contrary to Collective P-Acyclicity—better referred to, perhaps, as *Berthive* P-Acyclicity.

Now suppose Bertha *can* make intercriterial preferability comparisons. Let vertical distance in the above diagram represent degree of preferability. Then it is still eminently reasonable to suppose that x_1 $\mathrm{P}^v x_2$ $\mathrm{P}^v x_3$ $\mathrm{P}^v x_4$ $\mathrm{P}^v x_5$ $\mathrm{P}^v x_1$—especially if intercriterial preferability comparisons are numerically inspecific or at least somewhat imprecise, although I do not think this proviso really is necessary.

In the utility profile setup of section 4.3, we can even treat the relative *weight* or *importance* of criteria as a variable rather than a constant by interpreting each utility function as a measure of *adjusted criterial value*—criterial value multiplied by a coefficient of criterial importance.

Instead of calling $c_1, \ldots, c_n$ *criteria*, we can call them *factor, attributes, dimensions*, or *objectives*. We can even reinterpret them as mutually exclusive *eventualities*, or *states of nature*, making Bertha's problem one of *decision making under uncertainty*—or decision making under *risk*, if we reinterpret criterial importance as *probability*.

Preference cycles can reasonbly arise in *multidimensional* or *multifactor* choice situations of various sorts, individual as well as collective.

CHAPTER 5

HOW REASONABLE IS "RATIONALITY"?

5.1. PINNING BLAME

Deduce a contradiction from reasonable-looking, widely held assumptions, and you have a paradox; the better entrenched the assumptions, the more paradoxical the paradox. Some reasonable-looking assumption is not reasonable after all. Some widely held assumption is a widely made mistake. So it is with the assumptions involved in each impossibility theorem. All seem reasonable. All seem to be widely held. One, at least, is unreasonable, mistaken. Which one?

An assumption about a collective-choice process can be reasonable in a *positive* sense, in a *normative* sense, or both. It can correctly describe *real* choice processes, *reasonable* choice processes, or both. To be a reasonable assumption in either sense, a condition in *n, A, PROF*, and $\mathbb{P}$ need not correctly describe *all* real or reasonable collective-choice processes, so long as it correctly describes *some significant class* of real or reasonable collective-choice processes. Of the conditions before us, none is unreasonable *on its face* in either sense.

To be sure, two singly reasonable assumptions can be *conjointly* *un*reasonable: although each correctly described a significant class of real or reasonable choice processes, the two classes might be disjoint.

I shall divide all the conditions involved in the impossibility theorems of chapters 4 and 5 into four groups and examine the reasonableness of each group in turn.

I. The MINIMUM DEMOCRACY CONDITIONS require collective preferences (pairwise choices) to reflect individual preferences in specified ways. They include SMAC, Absolute Majority Condition, Virtual Unanimity, Positive Responsiveness, Pareto Preference, Strong Pareto Preference, Nondictatorship, Nonblocker, Weak Nonblocker, Nondespotism, Minimum Resoluteness, Weak Positive Responsiveness, Pareto Preference 2, Nondespotism 2, and Minimum Resoluteness 2.

The first four conditions are violated by many real collective-choice processes, and they are questionable as normative requirements in many circumstances. But the most interesting and general impossibility theorems do not involve these conditions. And by comparison, at least, the remaining conditions seem quite unobjectionable from both positive and normative points of view.

That will not stop the contentious minded, of course, from raising the following objections.

Objection 1. Although collective choices, properly so called, would perforce reflect individual preferences, at least to the limited degree postulated by the Minimum Democracy conditions, the impossibility theorems show that it is unreasonable ever to choose collectively—if the collective-choice idea even makes sense. *Governmental* choices need bear no particular relation to individual preferences. The proper business of government is not to satisfy people's preferences, but to serve their needs, promote their happiness, protect their welfare, guard their rights, fulfill their "real will," or some such thing. (Platonists, Hegelians, and Marxists, among others, might endorse this objection.)

Reply. The impossibility theorems do not require that we interpret P_i^v as Mr. *i*'s *preference* relation in the *v* situation. We can interpret P_i^v instead as the relation of x to y when, in the *v* situation, Mr. *i* needs x more than y, or x would make him happier than y, or x would protect his welfare better than y, or choosing y rather than x would violate some right of his, or his "real will" favors x over y, or whatever. And instead of interpreting N as an entire society, we can interpret it as a priesthood, the Vanguard of the Proletariat, or some such thing.

Besides, to reject collective choice per se as unreasonable or meaningless is itself too paradoxical to be a solution to any paradox.

Objection 2. From an *anarchist* point of view, collective choice per se is indeed unreasonable: social choices should never be made collectively; any alteration in the state of society should be the result of voluntary individual action or voluntary exchange.

Reply. A system of voluntary exchange is itself a collective-choice process, in the broad sense in which I have been using this phrase. The participants in a market economy collectively choose the prices of goods, the distribution of income among workers and producers, and the distribution of goods among consumers: these choices are attributable to all participants but to no one of them. A *collective*-choice process need not be highly centralized or coercive.

Objection 3. Suppose A comprises choice options that are no one's business but mine—options regarding the color of my underwear, for example. Then I ought to be a *dictator*, and in my society I am one.

Reply. Under the proffered interpretation of A, this set does not really comprise collective-choice options, except in a limiting sense. Under any normal governmental or collective-choice interpretation of N, A, *PROF*, and $\mathbb{P}$, Nondictatorship and its ilk are realistic and uncontroversial. These conditions are so mild, indeed, that the worst tyrannies in history never came close to violating them. Could Hitler have gotten his way on *all* possible pairwise choices in *all* possible situations in which literally *all* other Germans—the entire armed forces, government, and Nazi Party included—opposed him?

Objection 4. The *Pure Unanimity Rule* prescribes the choice of x over y when, and only when, everyone prefers x to y. This rule violates Minimum Resoluteness and Weak Positive Responsiveness. For it always requires collective indifference in the absence of unanimity.

Reply. The Pure Unanimity Rule is anomalous anyway: because rarely is either of two alternatives unanimously preferred to the other, Pure Unanimity Rule requires a coin toss (or some such thing) to settle virtually every choice. Collective-choice rules of this sort are, if nothing else, utterly unrealistic.

If someone found the Pure Unanimity Rule attractive, I suspect he would be confusing it with either the Conservative Unanimity Rule or the Condition of Pareto Optimality. Unlike the Pure Unanimity Rule, the Conservative variety requires unanimity, not in general, but only to overturn the status quo. And unlike the Pure Unanimity Rule, Pareto Optimality is not a collective-choice process at all, but a condition on collective-choice processes. It says that if x belongs to the choice set from α under the given collective-choice process, then there is no alternative in α that someone prefers and no one disprefers to x. It is the chief business of welfare economics to prove that market processes satisfy this condition in specified circumstances. Strong Pareto Preference is the restriction of Pareto Optimality to two-element feasible sets.

Objection 5. According to Minimum Resoluteness, Weak Positive Responsiveness, and the Paretian conditions, for any two alternatives in A there are circumstances under which those alternatives are not collectively indifferent: one is collectively preferred to the other. This implies that any two alternatives are *comparable* in the sense that their pair set is a potential feasible set. Although some pairwise collective-choice processes satisfy this universal-comparability condition, some do not. Procedures for choosing between candidates for offices satisfy it. Procedures for deciding on motions violate it: because every motion is pitted against the status quo, only pair sets containing a status-quo alternative are treated as potential feasible sets.

Reply. In the first place, even if universal comparability is not satisfied by all real or reasonable pairwise collective-choice processes, it is satisfied by a large and important class of such processes.

Furthermore, universal comparability is not violated nearly so often as the objection suggests. For one thing, parliamentary procedures usually allow motions to be compared, not only with status-quo alternatives, but with substitutes and amendments. And in most cases, a procedure that pits motions against status-quo alternatives is part of a sequential, or multistage, process in which the status quo at each successive stage is the alternative chosen at the previous stage. Since any alternative could have been chosen at the previous stage, depending on individuals' preferences and the order of voting, any two alternatives are comparable—they constitute a potential feasible set—under such a process.

II. The INDEPENDENCE CONDITIONS, which are special cases of Arrow's Independence of Irrelevant Alternatives, require collective pairwise choices to exhibit some measure of independence from individual preferences involving infeasible alternatives. They include Binary Independence, Weak Binary Independence, and Minimum Independence.

Arrow's paradox often is blamed on Binary Independence—or on the Independence of Irrelevant Alternatives. Those who take this line often contend, more specifically, that Binary Independence illicitly prevents collective choices from reflecting interpersonal comparisons of preference intensity.

We cannot solve the paradoxes of collective choice by declaring the Independence conditions unreasonable, for four reasons:

First, Binary Independence is widely satisfied—*universally* satisfied, as far as I can see—by extant collective-choice processes. What is more, such processes do not reflect interpersonal comparisons of preference intensity. So Binary Independence is unimpeachable as a *positive* assumption.

Second, the reasons we examined in section 2.3 for the *desirability* of violating Binary Independence were poor. And as I argued in section 2.2, it would not be reasonable, generally, to base collective choices on interpersonal comparisons of preference intensity, even if that were possible.

Third, Binary Independence is not essential to the paradoxes anyway. As you saw in section 4.2, we can replace Binary Independence with Weak Binary Independence and still prove an impossibility theorem very similar to Theorem 3.5.1. I take it that the likely objections to Binary Independence do not hold against Weak Binary Independence. And if we are willing to replace max(3, n)-Domain by Minimum Semiorder Condition, we can prove an impossibility theorem involving no independence condition at all—and very little else. Then there was the contradiction deduced in section 4.3 from apparently mild conditions that allowed (although they did not require) collective choices to be quite sensitive to interpersonal comparisons of preference intensity.

Fourth and most important, Binary Independence is just an *informational* restriction. It says collective choices are independent of certain information—information about individuals' preferences involving infeasible alternatives. Even if such information were *sometimes* available, or at least easily obtainable, it certainly is not *always* available or easily obtainable—or obtainable with difficulty, for that matter. So even if there were some reasonable collective-choice processes that used such information, there must also be some that do not, hence some that satisfy Binary Independence. To convict Binary Independence of unreasonableness in the normative sense is to hold, paradoxically, that there is no reasonable procedure for making collective choices in the absence of a type of information that often does not exist.

III. The DOMAIN CONDITIONS concern the membership of *N, A, PROF*, and *UPRF*. They include the assumptions that $n \geq 3$, $n \geq 5$, and $n \geq 8$, as well as 3-Domain, max(3, p + 2)-Domain, max(3, n)-Domain, γ-Domain, Discrete max(3, n)-Domain, Minimum Semiorder Condition, and Utility Condition.

These are less demanding and more realistic than the widely assumed Unrestricted Domain, as I explained in section 2.1.

As you know from section 4.1, we can avoid collective-preference cycles and other violations of Collective P-Transitivity by weakening the various k-Domain conditions so that they do not require a preference profile to belong to *PROF* unless it is single peaked or value restricted or Condorcet free or some such thing. But as you also know, these restrictions are not very realistic, whereas preference profiles that underlie collective-preference cycles are quite realistic. And applicability to realistic collective-choice situations is essential not only to real collective-choice processes but to reasonable ones.

Besides, what the Domain conditions require, in effect, is that the other conditions hold in a wide variety of hypothetical situations, some possibly less realistic than others; and it is common and proper in theoretical work to require an acceptable set of general assumptions to hold in comparatively unrealistic situations as well as more realistic ones. Philosophers require their definitions and normative principles to be immune even from bizarre, hypothetical counterexamples. Natural and behavioral scientists require their empirical hypotheses and models to be immune even from highly contrived experimental falsification. The reason is this: If a theoretical assertion yields no anomalous consequences even in unrealistic situations created for the purpose of refutation, it is all the more worthy of our confidence in realistic situations. So if a theoretical assertion yields anomalous consequences in unrealistic situations, we properly amend it to exorcise those consequences before trusting what the assertion tells us about more realistic situations.

The relation of a theoretical assertion T to possible situations is depicted in figure 5.1. We *test* T by looking for *clear* situations. These include *confirming* situations, in which what T says is clearly correct, and *disconfirming* situations, in which what T says is clearly incorrect;

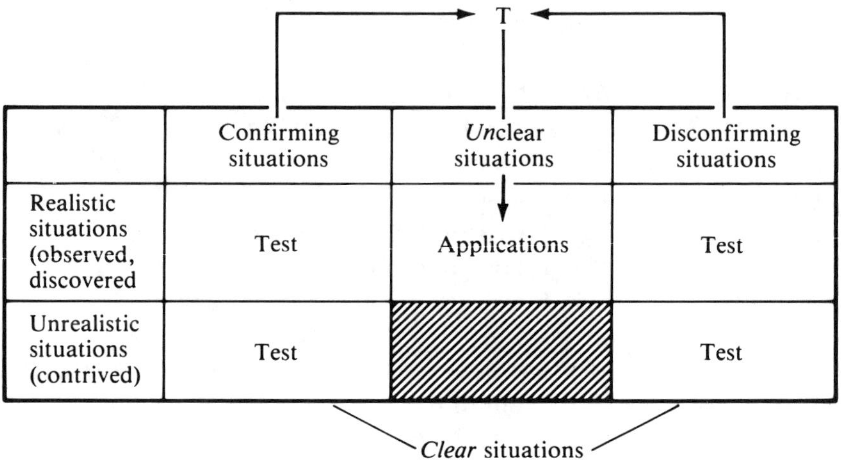

Figure 5.1

of course, if T is generally correct, there are no disconfirming situations. We expand the class of clear situations, and so get a better test, by including somewhat unrealistic situations. The likely objection to doing this—to requiring that T hold even in unrealistic situations—is that we need never *use* T to inform ourselves about such situations. But the same is true of the *clear realistic* situations. Both are used only for testing; they are no less important on that account.

Some theorists have faulted the Domain conditions, not for requiring that there be a collective choice in each of a wide variety of situations, but for requiring that these choices, although very differently situated, all issue from the same process—the one represented by $\mathbb{P}$. We ought to allow, they say, for the existence of different choice processes applicable to different situations, as opposed to one for all situations. Letting *PROF* satisfy your favorite Domain condition, they would partition *PROF* into a number of subsets, representing different kinds of situations, with a separate pairwise collective-choice process defined on each subset.

But even if this were done, there would perforce still exist a single pairwise collective-choice process defined for *all* $v \in PROF$. *Proof:* For $v \in PROF$, let F_v be the pairwise collective-choice process defined on that subset in the partition of *PROF* to which v belongs, hence defined for v, so that the value, $F_v(v)$, of F_v at v is the collective-preference relation in the v situation. Now define the function $\mathbb{P}$ on *PROF*:

$$\mathbb{P}^v \equiv F_v(v).$$

Then $\mathbb{P}$ represents a pairwise collective-choice process defined on the *whole* of *PROF*.

To be sure, $\mathbb{P}$ might behave in ways we intuitively regard as dissimilar when applied to preference profiles drawn from different sets in the partition of *PROF*. But the conditions used in the most general of our impossibility theorems allow this anyway.

Even if no F_v satisfies your favorite domain condition, $\mathbb{P}$ does. And so long as each F_v satisfies your favorite Minimum Democracy and Collective-Rationality conditions (Collective P-Acyclicity and the like), so must $\mathbb{P}$.

This is not necessarily true of the Independence conditions or Weak Positive Responsiveness. For each of these requires a certain profile-to-profile "consistency." So if it is not simply a logical error, the proposal we are examining is a round-about way of objecting to the Independence conditions or to Weak Positive Responsiveness. The latter itself embodies a good deal of Independence, and in the presence of Binary Independence it can be weakened so that it only requires the appropriate relationship to hold between very similar profiles. The main objection, I take it, is to Binary Independence. But this means the proposal in question requires collective choices to depend on a type of preferential information that often is not available.

The details are worth a close look: Suppose $v, w \in PROF$, $x, y \in A$, and $v/\{x, y\} = w/\{x, y\}$. But suppose $x\mathbb{P}^v y$ and not $x\mathbb{P}^w y$, contrary to Binary Independence, because F_v and F_w are very different processes. Then to choose between x and y, we must decide whether to invoke F_v or F_w, so we must decide whether we are in the v or the w situation. But these situations differ only in individual preferences regarding infeasible alternatives—alternatives in $A - \{x, y\}$. So to choose between x and y, we need information about individual preferences regarding infeasible alternatives—information that may well not be available.

IV. The COLLECTIVE-RATIONALITY CONDITIONS require that collective-preference relations have certain transitivitylike properties, notably acyclicity. These conditions include Collective P + I- and P-Transitivity; the latter plus any condition of the form (T) (discussed in section 3.4); Collective P-Acyclicity, P-Atricyclicity, and P-Acyclicity 2; and Weak Collective P-Acyclicity.

As the only remaining bones of contention, the Collective-Rationality conditions must be responsible for the paradoxes. Even the comparatively weak Collective P-Acyclicity, Collective P-Atricyclicity, and Weak Collective P-Acyclicity must be unreasonable assumptions, both normatively and positively.

But not surprisingly. Unlike the Independence and Domain conditions, the Collective-Rationality conditions are not *situational* conditions: they do not require that ℙ be applicable in certain situations, nor do they restrict the range of situations to which ℙ is applicable. And unlike the Minimum Democracy conditions, the Collective-Rationality conditions are not conditions of *personal treatment:* they do not require that people be treated in specified ways; more specifically, they do not require that people's interests or desires be respected—that they be treated as sovereign—to any degree. The Collective-Rationality conditions rather are abstract, mathematical theorists' assumptions. Whatever their technical utility of aesthetic appeal, they say nothing about circumstances of choice or choice-determining factors, nothing about people or their preferences. Unlike the other conditions, they are not forced upon us either by the constraints of reality or by the moral requirement to treat people in a minimally decent fashion.

As you will see in chapter 10, the traditional conditions of rationality ensure that choice processes have certain practical or "computational" advantages. But this is no compelling defense: the features in question are not essential to reasonable or realistic choice processes.

The Collective-Rationality conditions extend to collective choice a type of "rationality" customarily attributed to *individual* choice. Although perhaps not illegitimate on its face, this extension is not obviously legitimate, either. And as you saw in section 4.4, the classical "rationality" conditions are highly questionable even when attributed generally to individual choice.

From a positive point of view, the Collective-Rationality conditions are neither obviously reasonable nor obviously unreasonable. We know that real-world political institutions sometimes violate them. We do not know how often.

Note that the collective-rationality conditions are *general* assertions. They require $\mathbb{P}^v$ to be transitive or acyclic or some such thing for *every* $v \in PROF$. It is to this that I object, not to the requirement that $\mathbb{P}^v$ be transitive or whatever for *some* $v \in PROF$.

Some might take this view: Collective P-Acyclicity is *singly* reasonable, but it and some other singly reasonable conditions are *conjointly unreasonable*. Because the situational conditions require ℙ to be applicable to a wide variety of situations, whereas Collective P-Acyclicity requires ℙ to yield an acyclic relation in every situation to which it applies, drastically weakening the former can save the latter. Because Binary Independence requires that certain choices from one pair be invariant under changes in factors that determine choices from other pairs, whereas Collective P-Acyclicity requires a certain regularity among choices from various pairs, throwing out the former might save the latter. Because the Minimum Democracy conditions require pairwise choices to be single reasonable, whereas Collective P-Acyclicity requires that they relate to one another in a reasonable way, the two could conceivably be overly demanding when combined.

But because the situational conditions and conditions of personal treatment essential to an impossibility theorem involving Collective P-Acyclicity are so utterly weak as such conditions go, because they are satisfied by all or virtually all binary collective-choice processes found in the real world or capable of institutionalization in the real world, and because they *are* situational and personal-treatment conditions, the blame for any conjoint unreasonableness must lie with Collective P-Acyclicity.

5.2. "RATIONALITY" CONDITIONS AND THEIR RATIONALE

The classical "rationality" conditions formalize the thesis that *to choose rationally* is *to maximize* something—to choose in such a way that no other choice would have been better, or preferable, in some sense. The fundamental condition is:

BICH $C(\alpha) \equiv \{x \in \alpha \mid yPx \text{ for no } y \in a\}$,

where:

DEFINITION xPy *iff* $x, y \in A$, $x \neq y$, *and* $C(\{x, y\}) = \{x\}$.

Assuming Axioms C1 and C2, BICH implies:

P-ACYCLICITY *There are no P-cycles.*

I have argued that the impossibility theorems show P-Acyclicity to be unreasonable as a *general* assumption about collective preference—although not necessarily unreasonable when restricted to special situations. It seems unreasonable as a general assumption about individual preference as well.

I turn now to the reasons one might have for accepting BICH, and therewith P-Acyclicity, as a general assumption, positive or normative, about individual or collective preference and choice. I shall not bother with special reasons for postulating P-Transitivity or the stronger transitivity conditions.

Reason 1. The classical "rationality" conditions are intuitively appealing, attractive, plausible, or equivalent to conditions that are.

Criticism. In light of the impossibility theorems of chapter 3 and 4, this is a bad reason for two reasons:

First, we are logically shortsighted. Often we find a general statement intuitively appealing because we are unaware of its intuitively *un*appealing consequences. The impossibility theorems we have examined show that P-Acyclicity and its ilk have consequences to the effect that certain reasonable-looking conditions are not all satisfied. And these consequences are sufficiently unappealing, surely, to outweigh whatever intuitive appeal the "rationality" conditions themselves may have had.

Second, the intuitive appeal of the "rationality" conditions, however great, is of a sort that *counts for less* than the intuitive appeal of the conditions with which they conflict. Unlike all the other conditions involved in the impossibility theorems, the "rationality" conditions are neither *situational* conditions nor conditions of *personal treatment*. We find the other conditions intuitively appealing as normative assumptions because we want to be able to make collective choices in a wide, realistic range of situations and to avoid wronging people. We find them intuitively appealing as positive assumptions because it is so obvious that they correctly describe all or most extant collective-choice processes. The same cannot be said of the "rationality" conditions. They do not express facts about available information or other situational features (as the Independence conditions do), no one would take to the barricades in their behalf (as one might in behalf of Pareto Preference or Nondictatorship), and it is not at all obvious that they correctly describe *any* extant collective-choice processes.

Reason 2. Models of individual and collective choice that contain the traditional "rationality" conditions are logically fruitful and mathematically tractable. In effect, BICH defines C in terms of P, thereby reducing the problem of making, explaining, and evaluating choices generally to that of making, explaining, and evaluating *pairwise* choices. Maximization problems and transitive relations are well understood.

Criticism. Agreed. But the virtues of logical fruitfulness and mathematical tractability are no match for consistency. And as you will see, these virtues can be preserved, to a great extent, when BICH is replaced by weaker, paradox-free conditions.

Reason 3. If BICH holds, choices made according to C are *stable:* undominated under P, they *cannot be overturned* in favor of other choices—unless, of course, preferences or other choice-determining factors change. Such choices will not cycle endlessly.

Criticism. Stability is not obviously a virtue in all cases (exactly what is wrong with the continual overturning of choices?). It may even be a vice, if it encourages monopolization of political power. And if and when it is a virtue, it can be achieved in the face of P-cycles (or P-dominated members of choice sets) by institutional rules that forbid or make it hard to overturn collective choices once made. *Cycles* can exist without *cycling*, and cycling is not necessarily a Bad Thing.

What is more, even apart from institutional rules, a weak sort of stability can be had without BICH or P-Acyclicity, as you will see in chapter 7.

Reason 4. Maximization is equivalent to choosing as though one were pursuing some *end*, some *goal*, some *objective.*

Criticism. I have often heard the contention. I have never heard it made precise or proved, at least not without blatant question-begging.

Besides, in the example of section 4.4, it was the pursuit of goals or objectives that led to Bertha's preference cycle. True, Bertha pursued several goals, not one; she had to balance conflicting objectives against each other. But does that not often happen?

Reason 5. If, contrary to BICH, $C(\alpha)$ cannot always be equated with the set of P-undominated members of α, let us find a suitable relation other than P and postulate that $C(\alpha)$ always comprises just those members of α that are undominated under that other relation. This way we eschew P-Acyclicity while preserving the root idea underlying BICH, that to choose rationally is to maximize something—to choose a best alternative in some sense.

Criticism. As I pointed out in section 1.4, if ρ is *any* binary relation on A for which $C(\alpha)$ is identically the set of ρ-undominated members of α, then ρ must be P, defined as above. P is the *only* relation on A that can have this property.

Reason 6. So long as Axioms C1 and C2 hold, this is trivial:

(1) For every $\alpha \in S$, there is a binary relation ρ on A (a weak ordering, indeed) such that $C(\alpha)$ is the set of ρ-undominated members of α.

Proof: Let ρ be the relation of x to y when $x \in C(\alpha)$ and $y \in A - C(\alpha)$. Is (1) not all BICH requires of C?

Criticism. It is not. BICH requires this:

(2) There is a binary relation ρ on A such that, for every $\alpha \in S$, $C(\alpha)$ is the set of ρ-undominated members of α.

Although (1) follows from Axioms C1 and C2 alone, (2) does not. *Proof:* Let $A = \{1, 2, 3\}$, and define C on S by letting $C(\{1, 2\}) = \{1\}$ and $C(\alpha) = \alpha$ when $\alpha \neq \{1, 2\}$. Then as lately proved, C satisfies (1). But not (2): Were there a ρ such that, for every $\alpha \in S$, $C(\alpha)$ were the set of ρ-undominated members of α, we should have either $1\rho 2$ or $2\rho 2$ since $2 \notin C(\{1, 2\})$, yet neither $1\rho 2$ nor $2\rho 2$ since $2 \in C(\{1, 2, 3\})$.

But for the order of quantifiers ("there is . . . ρ . . . such that" and "for every $\alpha \in S$"), (1) and (2) are the same. Perhaps their superficial similarity explains why BICH strikes some as little more than a tautology.

Reason 7. Preference cycles can force choosers to behave in excessively costly ways. Davidson, McKinsey, and Suppes give this example: A department head offers Mr. S. three options: (a) a full professorship at \$5,000, (b) an associate professorship at \$5,500, (c) an assistant professorship at \$6,000. "Mr. S. reasons as follows: *aPb* since the advantage in kudos outweighs the small difference in salary; *bPc* for the same reason; *cPa* since the difference in salary is now enough to outweigh a matter of rank." This preference cycle can force Mr. S. into excessively costly behavior in the following way:

> The department head . . . says, "I see you prefer *b* to *c*, so I will let you have the associate professorship—for a small consideration. The difference must be worth something to you." Mr. S. agrees to slip the department head \$25 to get the preferred alternative. Now the department head says, "Since you prefer *a* to

b, I'm prepared—if you will pay me a little for my trouble—to let you have the full professorship." Mr. S. hands over another $25 and starts to walk away.... "Hold on," says the department head, "I just realized you'd rather have c than a. And I can arrange that—provided.... "

Criticism. Mr. S.'s preference for b over c does *not* force him or even make it reasonable for him to pay, however little, for b. For example, if he preferred b to c but a to both, it would be irrational for him to pay anything for b if he could get a for the same price.

It is reasonable for Mr. S. to pay a nominal fee for b *only* if $b \in C(\{a, b, c\})$ and he cannot get a member of $C(\{a, b, c\})$ without paying as much or more. But since $aPbPcPa$, it seems reasonable to suppose that $C(\{a, b, c\}) = \{a, b, c\}$: given a choice among a, b, and c, it is a matter of indifference to Mr. S. which he chooses (although the choice between any *two* of them would *not* be a matter of indifference). So Mr. S. does not have to bribe anyone to get a member of $C(\{a, b, c\})$. Whatever he gets is bound to belong to $C(\{a, b, c\})$. Hence, there is no reason for him to pay for b.

Let us complicate the example: Instead of offering Mr. S. the choice of a, b, or c all at once, the department head first (at time t_1) offers Mr. S. a choice of b or c, making a available only later, at t_2. Then, Davidson, McKinsey, and Suppes might contend, it is reasonable for Mr. S. to pay for b at t_1.

But as in the previous example, Mr. S will have no reason to pay a bribe at t_2. So if he acts according to his preferences, he will lose $25 at most. His predicament is not nearly so bad as Davidson et al. contend.

True, Mr. S.'s preferences might lead him at t_1 to spend money he would not have spent had he known a would become available at t_2. But *that is not because his preferences are cyclic.* When Mr. S.'s feasible set expands, lack of prescience concerning such expansion can cost him money. But that can happen *even with transitive preferences*, witness the following example:

As before, Mr. S.'s feasible set is $\{b, c\}$ at t_1 and $\{a, b, c\}$ at t_2, it costs him $25 at any time to ensure that he will get his choice, and he prefers a to b to c. But now he prefers a to c rather than c to a: having come into some money, he can afford transitive preferences. At t_1, then, Mr. S. spends $25 for b, just as in the last example. But at t_2, he spends another $25 for a. Had he known at t_1 that a would be feasible at t_2, he would have spent \$25 only at t_2. So just as in the last example, Mr. S.'s preferences plus his ignorance at t_1 of a's feasibility at t_2 cost him \$25 at t_1.

Let us complicate the example further: As before, Mr. S. is given a choice of b or c at t_1. But now his feasible set at t_2 is $\{a, b\}$ rather than $\{a, b, c\}$. Then it seems reasonable for Mr. S. to pay, not just for b at t_1, but for a at t_2. And if he is then offered the choice of c or a at t_3, of b or c at t_4, and so on, why should he not act as a money pump and pay for c at t_3, for b at t_4, and so on and on?

Although only a and b are feasible at t_2, Mr. S. will realize at t_2 that he is engaged in a drawn-out process of choosing among a, b, and c during an interval that includes t_1 and t_2. At t_2, there are just two feasible alternatives; during the longer interval, there are three. Relative to t_2, the feasible set is $\{b, c\}$; relative to the longer interval, it is $\{a, b, c\}$. At t_2, which set should Mr. S. treat as *the* feasible set? Should he spend his money at t_2 as if he were choosing between a and b? Or should he adopt a broader temporal perspective at t_2 and spend as if he were choosing among a, b, and c?

The latter approach is the more reasonable. If he adopts it, Mr. S. will not, after all, act as a money pump. He will simply behave at t_2 as he did in the previous example, spending nothing. Similarly for t_3, t_4, and so on.

In general, once you realize you are engaged in a drawn-out task of choosing among three alternatives, even though only two are feasible in any brief period, you should spend your money as if you are choosing among the three, paying for an alternative only if it belongs to the choice set from the set of all three alternatives and no other member of that choice set costs less.

I beg no question in averring this principle. For it does not just benefit choosers with cyclic preferences. Violations of it can be excessively costly *even when preferences are transitive*, as in this example:

As before, Mr. S.'s feasible set is $\{b, c\}$ at t_1 and $\{a, b\}$ at t_2, he prefers a to b and b to c, and it costs him \$25 at any time to ensure that he will get his choice. But now he knows at t_1 that a will be feasible at t_2, and his preferences are transitive: he prefers a to c rather than c to a. At t_1, Mr. S. would be crazy to spend his money as if he were choosing between b

and *c*, shelling out \$25 for the associate professorship (*b*). That would be money down the drain, and Mr. S. knows it. For he realizes he will end up with the full professorship (*a*) anyway, and it will cost him \$25 at t_2 whether or not he has already paid for *b*. The only reasonable strategy open to Mr. S. is to act on the proposed principle, spending \$25 just once—for *a* at t_2.

The main defect in the Davidson-McKinsey-Suppes argument, and in any argument designed to show that cyclic preferences beget excessively costly behavior, is one not of detail but of overall structure. We are told of Mr. S.'s bizarre situation and asked to assume *both* that his preferences are cyclic *and* that he acts according to certain prescribed *cost principles*—principles that tell one what costs one should be willing to incur, given one's preferences. From this it is deduced that Mr. S. is willing to incur what apparently are excessive costs. Davidson et al. blame this on Mr. S.'s cyclic preferences. But it is equally open to us to pin the blame, as I did, on the imputed cost principles. This is not just an acceptable tack, but the only reasonable one, in view of the unacceptable consequences of P-Acyclicity.

The cost principles Davidson et al. implicitly assumed were refutable even in the presence of strictly transitive preferences. I suspect this is the case with *any* argument designed to show that preference cycles force choosers to incur excessive costs, but I cannot prove it.

SOURCES AND RELATED CONTRIBUTIONS

As far as we know, Condorcet (1785) was first to report the Classical Voting Paradox. Discussed from time to time in the nineteenth century, this phenonemon was largely ignored in the twentieth until Black (1948) and Arrow (1952) focused attention on it. See Black (1958: Part II) for the early history of the subject.

Apart from the names of some conditions, the version of Arrow's Theorem presented here comes from Arrow (1963: ch. 8). This differs from the earlier version of Arrow (1950; 1952: ch. 5; 1963: ch. 5) in two respects. First, in place of Pareto Preference, the earlier version used two other conditions:

POSITIVE ASSOCIATION OF SOCIAL AND INDIVIDUAL VALUES *Suppose*
$i \in N$,
$x, y \in A$,
$v, w \in PROF$,
$P_j^v = P_j^w$ *for all* $j \neq i$ *in* N,
$P_i^v/(A - \{x\}) = P_i^w/(A - \{x\})$, *and*
for all z, $zP_i^w x$ *only if* $zP_i^v x$.
Then $x\mathbf{P}^w y$ *if* $x\mathbf{P}^v y$, *and* $y\mathbf{P}^v x$ *if* $y\mathbf{P}^w x$.

CITIZENS' SOVEREIGNTY *For all distinct* $x, y \in A$, $x\mathbf{P}^v y$ *for some* v.

It is easy to see that these conditions together imply Pareto Preference. Second, in place of Unrestricted Domain, the original version used this condition:

Every $v \in PROF$ comprises only weak orderings of A, and some three-member subset of A is completely free.

Blau (1957) showed that this original version of Arrow's "Theorem" is false but becomes true on the addition of either Unrestricted Domain or the following domain condition:

Every three-member subset of A is completely free.

As you know, 3-Domain is weaker than either of these conditions: it is enough to demand that every three-member subset of A be *linearly* free. Murakami (1961a) left Arrow's original condition alone and proved the theorem after weakening Nondictatorship.

For further discussion of Arrow's Theorem, see Luce and Raiffa (1957: ch. 14), Vickrey (1960), Arrow (1967a), Sen (1970: chs. 3, 3*, 4, and 4*), Wilson (1972), Schwartz (1970 and 1973: ch. 8), Fishburn (1973: chs. 14–17), Plott (1976), and Kelley (1978: chs. 1–5). Plott's widely admired paper is, I think, the most illuminating presentation. Hansson (1969) proves an equivalent theorem for a *fixed* feasible set, using a condition that combines Independence of Irrelevant Alternatives with the Weak Axiom of Revealed Preference, an equivalent of the classical rationality conditions (see section 10.2 below).

Theorem 3.3.1, the oligarchy theorem, was discovered independently by Allan Gibbard and me in 1968–69, but neither of us published it. (I had to excise a half-page footnote on it from Schwartz [1970] owing to length limits.) This result first appeared in Mas Colell and Sonnenschein (1972). See also Fishburn (1973: ch. 16) and Kelley (1978: ch. 4). Meanwhile, I included the oligarchy theorem in various versions of my classnotes, *Notes on the Abstract Theory of Collective Choice*, at Stanford and Carnegie-Mellon from 1970 onward.

Theorem 3.4.1, based on transitivity conditions between Collective P + I-Transitivity and P-Transitivity, generalizes the theorem of Blair and Pollack (1979) and Blau (1979), which is based on transitivity conditions of the form (2) in Lemma 3.4.1. That result in turn generalizes a version of Arrow's Theorem based on collective PIP-Transitivity, given in my *Notes*.

Theorem 3.5.1 is a special case of Theorem 4.3.1—the impossibility theorem that allows some use of interpersonal utility comparisons—which comes from Schwartz (1970) and is the first Arrow-style impossibility theorem based on Collective P-Acyclicity (or, indeed, on any weakening of Collective P + I-Transitivity). We can turn Theorem 4.3.1 into Theorem 3.5.1, in effect, by adding Binary Independence and the assumption of ordinality of individual utility ($\mathbb{B}$ is invariant under arbitrary monotonic transformations of its arguments). Theorem 4.2.1, which uses preference relations rather than utility measures but still has a weakened independence condition, comes from my dissertation (University of Pittsburgh, Department of Philosophy, 1969). It, too, is a special case of Theorem 4.3.1. I did not previously publish Theorem 3.5.1 or 4.2.1 because of the greater generality of Theorem 4.3.1.

Theorem 3.5.2 is similar to that of Mas Colell and Sonnenschein (1972), as modified by Blair et al. (1976) to include Collective P-Atricyclicity in place of Collective P-Acyclicity. There are two differences between my theorem and theirs: First, where they assume $n \geq 4$, I assume $n \geq 8$. Second, where they assume Unrestricted Domain, I assume 3-Domain. The second change required a different proof strategy: Mas Colell and Sonnenschein make essential use of profiles containing nonlinear orderings. Theorem 3.6.1, which obviously generalizes Theorem 3.5.2, comes from Schwartz (1982c).

Theorems 4.2.1 and 4.3.1 both use weakened independence conditions, and Theorem 4.3.1 (from Schwartz 1970) explicitly allows collective choices to reflect interpersonal utility comparisons to a great extent. Waldner (1973) contested the significance of this theorem on the basis of the contention that I somehow *assumed interpersonal comparability*. I did no such thing: I merely avoided the contrary assumption. For attempts to devise collective-choice rules based on interpersonal utility comparisons, see Goodman and Markowitz (1952) and Sen (1970). Samuelson (1967) conjectured an impossibility theorem based on cardinal *but not interpersonal* utility comparisons. Campbell (1973) and Kalai and Schmiedler (1977) proved such theorems. Needless to say, my 1970 result had already generalized Samuelson's conjecture.

Theorem 4.2.2, which allows individuals to have semiorders that are not weak orders, comes from Packard (1975). Fishburn (1970b, c) also discusses collective choice in the presence of nontransitive individual indifference.

Black (1948, 1958) formulated the Single Peakedness condition and the "median-voter theorem." The root idea comes from Hotelling (1929), whose main concern was not voting but certain economic applications, such as the optimal location of a business. Downs (1957) used the same basic assumption, now called the one-dimensional spatial voting model, to explain electoral competition: In a two-candidate election that satisfies the model, the candidate closer on the single dimension (or horizontal axis) to the median of the voter's ideal points is preferred to the other candidate by a majority, so candidates will try to adopt positions close to the median.

The spatial theory of voting has become a major academic industry. The first contribution to multidimensional spatial theory is that of Black and Newing (1951), which was ahead of its time and has not enjoyed that recognition it deserves. Other major pioneer contributions are those of Black (1958: 130–139), Davis and Hinich (1966), Plott (1967), and Tullock (1968). For an authoritative, up-to-date exposition, see Enelow and Hinich (1984). The theorem about a median in all directions comes from Davis, DeGroot, and Hinich (1972). McKelvey (1976), Schofield (1978), and Cohen (1979) have, in effect, generalized the three-voter instability example given in section 4.1, proving the near inevitability of a universal majority-preference cycle in a space of two or more dimensions. Their assumptions, however, are somewhat restrictive, and when these are replaced by natural alternative assumptions, we get an analytically fruitful Black-like stability result for two dimensions (although not for three or more), as shown by McCubbins and Schwartz (1985).

Quite a few restrictions on combinations of preferences have been shown to guarantee Collective P-Transitivity (various transitivity conditions, in fact) under majority rule. The main results are those of Black (1948, 1958), Arrow (1952 and 1963: ch. 7), Sen and Pattanaik (1969), Inada (1969), Sen (1970: chs. 10 and 10*), Pattanaik (1971), and Fishburn (1973: chs. 8–11). The conditions of Condorcet Freedom and Condorcet Effectiveness have not previously appeared in print. Neither has the generalization from Simple Majority Rule to arbitrary voting rules satisfying Independence, Strong Pareto Preference, and Overlap.

The prevalence of nontransitive or cyclic collective preferences or of unstable collective choices has been investigated by three other approaches not discussed in Part II. One approach that has received insufficient attention is *empirical*. Riker (1958), Blydenberg (1971), Sullivan (1976), and Dobra (1983) all present empirical evidence of the "paradox of voting." A second approach seeks to reckon the *probability* of majority-preference cycles, given a uniform probability distribution over preference profiles; see Garmen and Kamien (1968), Niemi and Weisberg (1968), and DeMeyer and Plott (1970). The third approach, pursued in detail in chapter 11, is to show that very mild and realistic assumptions couched in behavioral or institutional terms logically guarantee instability.

The interpretation of impossibility theorems in terms of *individual* choice is discussed by May (1954), Schwartz (1970, 1972), and Packard (1975a,b; and 1982).

Different approaches to solving or dissolving the paradoxes of collective choice may be found in Arrow (1963: ch. 8), Luce and Raiffa (1957: ch. 14), Sen (1970), Plott (1976), and Abrams (1980).

Davidson, McKinsey, and Suppes (1954), Tullock (1964), Tversky (1969), and Samuelson (1972) all defend the traditional assumptions of rational choice. Davidson et al. and Tverski both offer the money-pump argument condemned in section 5.2. Schwartz (1970, 1972) argues against BICH and P-Acyclicity. May (1954) and Tversky (1969) offer empirical evidence against the traditional rationality assumptions, viewed as hypotheses about individual behavior. (Tversky still regards violations of these assumptions as anomalous.) The responsibility of the collective-rationality assumptions for the impossibility theorems is discussed by Buchanan (1954), Tullock (Appendix 2 of Buchanan and Tullock 1962), Fishburn (1970e), Schwartz (1970, 1971), and Plott (1976).

PART III

THE GENERAL THEORY OF SOLUTIONS

CHAPTER 6

GETCHA AND GOCHA

6.1. SPAYING BICH

You can phrase BICH this way:

$C(\alpha) \equiv$ the set of P-undominated elements of α,

where:

DEFINITION *x is a ρ-undominated element of α iff $x \in \alpha$, ρ is a binary relation, and nothing in α bears ρ to x.*

(Notice that this defines undomination for *arbitrary* binary relations, not just for P. Subsequent definitions conform to the same policy.)

A virtue of BICH is that it defines C, in effect, in terms of P, thereby reducing the problem of making, explaining, and evaluating choices generally to that of making, explaining, and evaluating *pairwise* choices. More generally, BICH gives a simple, intuitive specification of the *choice set* from any potential feasible set. When collective-choice problems are thought of as games, the same point can be made this way: BICH formulates a simple, intuitive *solution* concept, the solution of a game (or collective-choice problem) being the set of possible outcomes that could obtain if participants acted rationally, given their preferences.

BICH also has a devastating drawback. Assuming Axioms C1 and C2, BICH implies:

every $\alpha \in S$ has at least one P-undominated element,

which is equivalent to P-Acyclicity, and I need not remind you what a Bad Thing *that* is.

Let us now see if we can weaken BICH so that its virtue is preserved and its drawback dropped. I shall entertain two proposals for doing this. I advocate neither as a general characterization of choice sets. From a normative point of view, both have important shortcomings. From a positive point of view, both are applicable in important collective-choice contexts, as you will see in chapters 9 and 10.

GETCHA

Suppose that $P/\{x, y, z, t\}$ has this form (where "... → ____" means "...P____"):

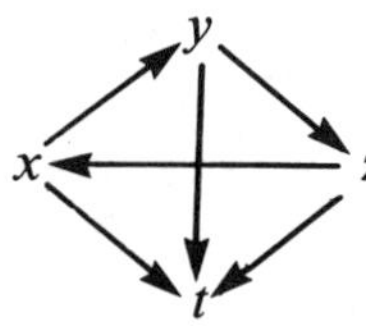

Then $\{x, y, z\}$ is a P-cycle, and $\{x, y, z, t\}$ has no P-undominated element.

But it does have a *P-dominant subset*, where:

> DEFINITION β *is a* ρ-dominant subset of α *iff* β *is a nonempty subset of* α, ρ *is an asymmetric binary relation, and* $x\rho y$ for every $x \in \beta$ *and* $y \in \alpha - \beta$.

Specifically, $\{x, y, z\}$ is a P-dominant subset of $\{x, y, z, t\}$.

If α has no P-undominated element but does have a P-dominant subset, ought we to equate $C(\alpha)$ with that subset? Not necessarily. Every set is a P-dominant subset of itself, after all: everything in α bears P to everything in $\alpha - \alpha$, there being nothing in $\alpha - \alpha$.

What distinguishes $\{x, y, z\}$ in the example is not that it is a P-dominant subset of $\{x, y, z, t\}$ but that it is a *minimum* one.

> DEFINITION β *is a* minimum ρ*-dominant subset of* α *iff* β *is a* ρ*-dominant subset of* α *and no proper subset of* β *also is a* ρ*-dominant subset of* α.

And every set belonging to S does have a unique minimum P-dominant subset. More generally:

> THEOREM 6.1.1. *If* ρ *is any asymmetric binary relation then every finite, nonempty set* α *has a unique minimum* ρ*-dominant subset.*

Proof. α is a ρ-dominant subset of itself. Thus, since α is finite, α has at least one *minimum* ρ-dominant subset. For uniqueness, suppose β and γ are minimum ρ-dominant subsets of α; to prove that $\beta = \gamma$. If neither $\beta \subseteq \gamma$ nor $\gamma \subseteq \beta$, there would be an $x \in \beta - \gamma$ and a $y \in \gamma - \beta$, whence $x\rho y$ and $y\rho x$, contrary to the asymmetry of ρ. Thus, either $\beta \subseteq \gamma$ or $\gamma \subseteq \beta$. But because β and γ are *minimum* ρ-dominant subsets of α, neither $\beta \subset \gamma$ nor $\gamma \subset \beta$. Hence, $\beta = \gamma$. ■

This theorem suggests one of the two proposals advertised earlier—that BICH be replaced by:

GENERALIZED TOP-CHOICE ASSUMPTION (GETCHA) $C(\alpha) \equiv$ *the minimum P-dominant subset of* α.

Like BICH, GETCHA defines C in terms of P, thereby reducing the problem of making, explaining, and evaluating choices generally to that of making, explaining, and evaluating *pairwise* choices.

Unlike BICH, GETCHA guarantees that $C(\alpha)$ never is empty, independently of any properties of P other than asymmetry. This follows immediately from Theorem 6.1.1, since by definition a P-dominant subset must be nonempty.

Like BICH, GETCHA implies my definition of P as the relation of x to y when $x, y \in A$, $x \neq y$, and $C(\{x, y\}) = \{x\}$, assuming only that P is an asymmetric binary relation on A.

THEOREM 6.1.2 *Let* ρ *be any asymmetric binary relation on A such that*

$C(\alpha) \equiv$ *the minimum* ρ*-dominant subset of* α.

Then $x\rho y$ *iff* $x, y \in A$, $x \neq y$, *and* $C(\{x, y\} = \{x\}$.

Proof. Suppose $x\rho y$. Then by hypothesis of the Theorem, $x, y \in A$ and not $y\rho x$, so $x \neq y$ and $\{x\}$ is the minimum ρ-dominant subset of $\{x, y\}$, whence $C(\{x, y\}) = \{x\}$. Conversely, suppose $x, y \in A$, $x \neq y$, and $C(\{x, y\}) = \{x\}$. Then $\{x\}$ is the minimum ρ-dominant subset of $\{x, y\}$, so $x\rho y$. ■

GOCHA

To motivate the second proposal, suppose $P/\{x, y, z, t, u, s\}$ has this form (where the absence of an arrow symbolizes indifference):

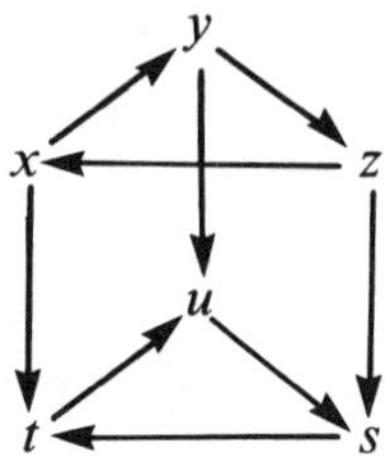

Then there are two P-cycles, $\{x, y, z\}$ and $\{t, u, s\}$, and $\{x, y, z, t, u, s\}$ has no P-undominated element. It has a minimum P-dominant subset, of course. But that subset is $\{x, y, z, u, s\}$ itself. The reason it is not $\{x, y, z\}$ is that *not every* element of $\{x, y, z\}$ bears P to *every* element of $\{t, u, s\}$; x, for example, does not bear P to u.

Although $\{x, y, z\}$ is not a minimum P-dominant subset of $\{x, y, z, t, u, s\}$, it *is* a minimum *P-undominated* subset, where:

> DEFINITION β *is a* ρ-undominated subset *of* α *iff* β *is a nonempty subset of* α, ρ *is a binary relation, and nothing in* $\alpha - \beta$ *bears* ρ *to anything in* β,

and:

> DEFINITION β *is a minimum* ρ*-undenominated subset of* α *iff* β *is a* ρ*-undominated subset of* α, *and no proper subset of* β *is a* ρ*-undominated subset of* α.

But we cannot identify $C(\alpha)$ in all cases with the minimum P-undominated subset of α, for α might have *more than one* such subset. Suppose that $P/\alpha = \{x, y, z, w, s, t, u, v, x', y', z', r\}$ has this form:

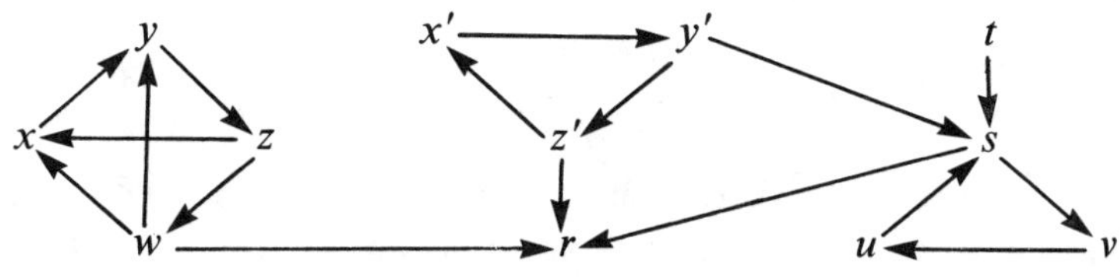

Then α has three minimum P-undominated subsets, $\{x, y, z, w\}$, $\{x', y', z'\}$, and $\{t\}$.

What we can do is to equate $C(\alpha)$ in each case with the *union* of minimum P-undominated subsets of α:

> GENERALIZED OPTIMAL-CHOICE AXIOM (GOCHA) $C(\alpha) \equiv$ *the union of minimum P-undominated subsets of* α.

Like BICH and GETCHA, GOCHA defines C in terms of P.

Unlike BICH, GOCHA guarantees that $C(\alpha)$ never is empty, independently of any properties of P, even asymmetry. This is because, for any binary relation ρ, every set is a ρ-undominated subset of itself, so every finite set has at least one *minimum* ρ-undominated subset, which by definition must be nonempty.

Like BICH and GETCHA, GOCHA implies my definition of P, assuming that P is an asymmetric binary relation on A.

THEOREM 6.1.3. *Let* ρ *be any asymmetric binary relation on* A *such that*

$C(\alpha) \equiv$ *the union of minimum* ρ*-undominated subsets of* α.

Then $x\rho y$ *iff* $x, y \in A$, $x \neq y$, *and* $C(\{x, y\}) = \{x\}$.

Proof. Suppose $x\rho y$. Then $x, y \in A$ and, by asymmetry, not $y\rho x$, whence $x \neq y$ and $\{x\}$ is the one and only minimum ρ-undominated subset of $\{x, y\}$, so that $C(\{x, y\}) = \{x\}$. Conversely, if $x, y \in A$, $x \neq y$, and $C(\{x, y\}) = \{x\}$, then $\{y\}$ cannot be a ρ-undominated subset of $\{x, y\}$, so $x\rho y$. ■

Dominant and Undominated Subsets

GOCHA and GETCHA can be given formally similar renditions if, in addition to P, we use the relation R, defined thus:

DEFINITION $R = A^2 - P^{-1}$ (*in other words,* xRy *iff* $x, y \in A$ *and not* yPx).

This is equivalent to

$$R = P \cup I.$$

So R is the relation of *preference or indifference*, sometimes called *weak preference*.

Among the elements of A, dominance under P is the same as nondominance under R, and nondominance under P is the same as dominance under R. So we can recast GETCHA in terms of R-undominated subsets, and GOCHA in terms of R-dominant subsets.

THEOREM 6.1.4. *Suppose* ρ *is an asymmetric binary relation on* A, $\rho' = A^2 - \rho^{-1}$, *and* $\alpha \in S$. *Then:*

(a) β *is a* ρ'*-undominated subset of* α *iff* β *is a* ρ*-dominant subset of* α.

(b) β *is a* ρ'*-dominant subset of* α *iff* β *is a* ρ*-undominated subset of* α.

(c) *These sets are identical:*
the minimum ρ*-dominant subset of* α,
the minimum ρ'*-undominated subset of* α,
the union of minimum ρ'*-undominated subsets of* α.

(d) *These sets are identical:*
the union of minimum ρ*-undominated subsets of* α,
the union of minimum ρ'*-dominant subsets of* α.

(e) *The following three conditions are equivalent:*
$C(\alpha) \equiv$ *the minimum P-dominant subset of* α (*GETCHA*).
$C(\alpha) \equiv$ *the minimum R-undominated subset of* α.
$C(\alpha) \equiv$ *the union of minimum R-undominated subsets of* α.

(f) *These two conditions also are equivalent:*
$C(\alpha) \equiv$ *the union of minimum P-undominated subsets of* α (*GOCHA*).
$C(\alpha) \equiv$ *the union of minimum R-dominant subsets of* α.

Proof. (a) and (b) follow immediately from the hypothesis. (c) and (d) follow immediately from (a) and (b), inasmuch as the unique minimum ρ'-dominant subset of α also is the *union* of such subsets. And (e) and (f) follow immediately from (c) and (d). ∎

6.2. EQUIVALENT FORMULATIONS: TOP CYCLES AND ANCESTRALS

GOCHA equates $C(\alpha)$ with the union of minimum P-undominated subsets of α. Among these subsets are all sets of P-undominated elements of α. Among them also are all *top P-cycles* in α.

DEFINITION β *is a* top ρ-cycle in α *iff* β *is a* ρ*-cycle* (*for some k and* $x_1, \ldots, x_k$, $\beta = \{x_1, \ldots, x_k\}$ *and* $x_1 \rho x_2 \cdots \rho x_k \rho x_1$), *and* β *is a* ρ*-undominated subset of* α ($\beta \subseteq \alpha$ *and nothing in* $\alpha - \beta$ *bears* ρ *to anything in* β).

As a matter of fact, *every* minimum P-undominated subset of a set $\alpha \in S$ must be either the unit set of some P-undominated member of α or else a top P-cycle in α. More generally:

THEOREM 6.2.1. *β is a minimum ρ-undominated subset of α iff β is either the unit set of a ρ-undominated element of α or else a top ρ-cycle in α.*

Proof. Suppose β is the unit set of a ρ-undominated element of α. Then β is a ρ-undominated subset of α. It is a *minimum* one because it is a unit set.

Now suppose β is a top ρ-cycle in α, hence a ρ-undominated subset of α; to prove that no proper subset of β also is a ρ-undominated subset of α. Suppose λ were a proper subset of β and a ρ-undominated subset of α; to deduce a contradiction. Then λ and β − λ are both nonempty; say $x \in \beta - \lambda$ and $y \in \lambda$. Since β is a ρ-cycle, there are $z_1, \ldots, z_k \in \beta$ such that $z_1 = x \in \beta - \lambda$, $z_k = y \in \lambda$, and $z_i \rho z_{i+1}$, $i = 1, 2, \ldots, k - 1$. Hence, there is an i for which $z_i \in \beta - \lambda$, $z_{i+1} \in \lambda$, and $z_i \rho z_{i+1}$, contrary to the ρ-undominated character of λ.

Suppose, conversely, that β is a minimum ρ-undominated subset of α; to prove that β is either the unit set of a ρ-undominated element of α or else a top ρ-cycle in α. Suppose β is not the unit set of a ρ-undominated element of α. Then since β is a ρ-undominated subset of α, to show that β is a top ρ-cycle in α it suffices to show that β is a ρ-cycle. Let $x, y \in \beta$ and $x \neq y$; it suffices to show that there are $z_1, \ldots, z_k \in \beta$ such that $z_1 = x$, $z_k = y$, and $z_i \rho z_{i+1}$, $i = 1, 2, \ldots, k - 1$.

Let β* be the least set μ containing y and containing everything in β that bears ρ to any element of μ. By construction, for every $w \neq y$ in β*, there are $z_1, \ldots, z_k \in \beta^*$ such that $w = z_1 \rho z_0 \rho \cdots \rho z_k \rho y$. So it suffices to show that β* exhausts β.

By construction, whatever in β bears ρ to anything in β* itself belongs to β*. So nothing in β − β* bears ρ to anything in β*. Hence, since β is a ρ-undominated subset of α, so is β*. Therefore, since β is a *minimum* one, β* cannot be a *proper* subset of β. So β* = β. ■

This theorem plus Theorem 6.1.4(e) immediately yield an alternative formulation not only of GOCHA but of GETCHA.

COROLLARY 6.2.1 (a) *For every $\alpha \in S$, β is a minimum P-undominated subset of α iff β is either the unit set of a P-undominated element of α or else a top P-cycle in α. Hence, GOCHA is equivalent to:*

C(α) ≡ the union of all unit sets of P-undominated elements of α and all top P-cycles in α.

(b) *For every* $\alpha \in S$, *the minimum P-dominant subset of* α *is either the unit set of the R-undominated element of* α *or else the top R-cycle in* α. *Hence, every* $\alpha \in S$ *has either a unique R-undominated element (which bears P to every other element) or else a unique top R-cycle, but not both. Hence also, GETCHA is equivalent to:*

$C(\alpha) \equiv$ *either the unit set of the R-undominated element of* α *or else the top R-cycle in* α.

GOCHA and GETCHA admit of another interesting type of reformulation. The *ancestral* of a binary relation ρ—also called the *asymmetric transitive closure* of ρ—is the result of transitizing ρ, then taking the asymmetric factor of the relation thus obtained. It stands to ρ as the ancestor-descendant relation stands to the parent-offspring relation. Someone is your ancestor if there is a parent-offspring chain running from him to you.

DEFINITION σ *is a* ρ-chain *from x to y iff* ρ *is a binary relation, and for some k there are* $x_1, \ldots, x_k$ *such that* $x_1 = x$, $x_k = y$, $\sigma = (x_1, \ldots, x_k)$, *and* $x_1 \rho x_2 \rho \cdots \rho x_k$.

DEFINITION T *is the* ancestral *of* ρ *iff* T *is the relation* $\{(x, y) \mid$ *there is a* ρ*-chain from x to y but none from y to x*$\}$.

THEOREM 6.2.2. *Let* ρ *be a binary relation,* α *a finite set, and* ρ^*_α *the ancestral of* ρ/α. *Then x belongs to some minimum* ρ*-undominated subset of* α *(hence to the union of such subsets) iff x is a* ρ^*_α*-undominated element of* α.

Proof. Suppose x belongs to a minimum ρ-undominated subset β of α. By Theorem 6.2.1, either x is a ρ-undominated element of α and $\beta = \{x\}$, or else β is a top ρ-cycle in α. In the first case, x also is a ρ^*_α-undominated element of α. If, instead, β is a top ρ-cycle in α, then nothing in $\alpha - \beta$ bears ρ to x, whence nothing in $\alpha - \beta$ bears ρ^*_α to x. And for every $y \in$ b, since β is a ρ-cycle and $\beta \subseteq \alpha$, there is a ρ/α-chain from x to y, so that not $y\rho^*_\alpha x$. Hence, nothing in α bears ρ^*_α to x if β is a top ρ-cycle in α. Either way, then, x is a ρ^*_α-undominated element of α.

Conversely, suppose x is a ρ^*_α-undominated element of α; to prove that x belongs to some minimum ρ-undominated subset of α. Let $\mu = \{x\}$ if x belongs to no ρ/α-cycle; otherwise, let μ be a ρ/α-cycle containing x that is *maximal* in the sense that it is not a proper subset of any other ρ/α-cycle. By Theorem 6.2.1, it suffices to show that μ is a

ρ-undominated subset of α. Suppose, on the contrary, that $y\rho z$ for some $y \in \alpha - \mu$ and $z \in \mu$; to deduce a contradiction. Since μ is either $\{x\}$ or a ρ/α-cycle, there is a ρ/α-chain from y to x. Were there also a ρ/α-chain from x to y, $\mu \cup \{y\}$ would be a ρ/α-cycle, contrary to the maximal character of μ. Therefore, there is no such ρ/α-chain. Hence, $y\rho^*_\alpha x$, contrary to the assumption that x is a ρ^*_α-undominated element of α. ■

Combined with Theorem 6.1.4(e), Theorem 6.2.2 immediately yields:

COROLLARY 6.2.2. (a) *If* $\alpha \in S$ *and* P^*_α *is the ancestral of* P/α, *then the union of minimum P-undominated subsets of* α *is the set of* P^*_α-*undominated elements of* α. *Hence, GOCHA is equivalent to:*

$C(\alpha) \equiv \{x \mid x$ *is an undominated element of* α *under the ancestral of* $P/\alpha\}$.

(b) *If* $\alpha \in S$ *and* R^*_α *is the ancestral of* R/α, *then the minimum P-dominant subset of* α is the set of R^*_α-*undominated elements of* α. *Hence, GETCHA is equivalent to:*

$C(\alpha) \equiv \{x \mid x$ *is an undominated element of* α *under the ancestral of* $R/\alpha\}$.

6.3. BICH, GETCHA, AND GOCHA: THEIR CONNECTIONS

Corollaries 6.2.1 and 6.2.2 help clarify the connection among the choice sets specified by BICH, GETCHA, and GOCHA.

According to Corollary 6.2.1, the BICH set always is a subset of the GOCHA set: if $\alpha \in S$, every P-undominated element of α belongs to the union of minimum P-undominated subsets of α.

But if there are P-cycles, of course, the GOCHA set is not always a subset of the BICH set. They are the same set only when P is acyclic. And they are the same *whenever* P is acyclic.

THEOREM 6.3.1. *P-Acyclicity holds iff, for every* $\alpha \in S$, $\{x \in \alpha \mid yPx$ *for no* $y \in \alpha\}$ = *the union of minimum P-undominated subsets of* α. *Hence, P-Acyclicity holds iff BICH and GOCHA are equivalent.*

Proof. Assume P-Acyclicity. Then by Corollary 6.2.1, all and only unit sets of P-undominated elements of α are minimum P-undominated subsets of α. Suppose, on the other hand, that P-Acyclicity does not hold. Then there is a P-cycle, α. By Corollary 6.2.1, the union of minimum P-undominated subsets of α is a α itself, whereas $\{x \in \alpha \mid yPx$ for no $y \in \alpha\} = \emptyset \neq \alpha$. ■

What about the GOCHA and GETCHA sets? Suppose that $P/\alpha = \{x, y, z, w, v, t\}$ has the following form:

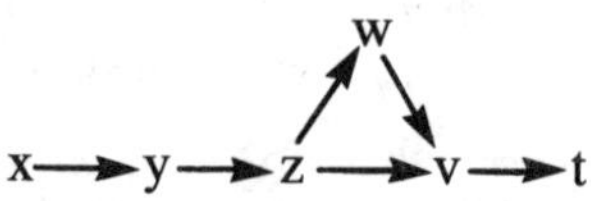

Then $\{x\}$ is the one and only minimum *P-undominated* subset of α. But the minimum *P-dominant* subset of α is α itself. The same is true even for the example shown in figure 6.1. To exclude t, at least, from the minimum P-dominant subset of α, we should need an arrow running directly from the designation of every other alternative in α to that of t—a stiff requirement.

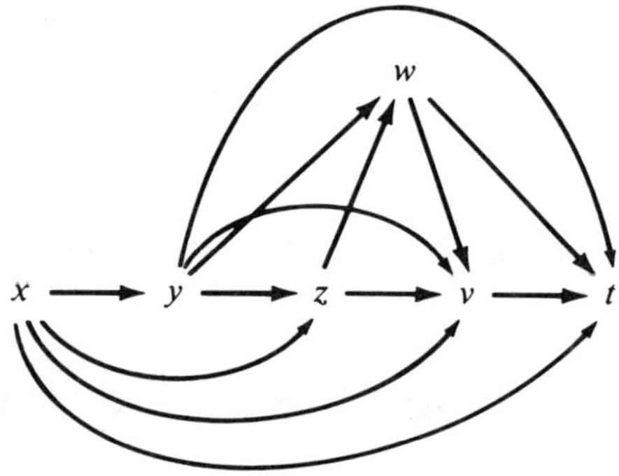

Figure 6.1

This example suggests that the GETCHA set is typically more inclusive than the GOCHA set. The former is perforce a superset of the latter, as I shall now prove. It can be a proper superset, as you just saw.

THEOREM 6.3.2. *Suppose* $\alpha \in S$, β *is a minimum P-undominated subset of* α, *and* δ *is the minimum P-dominant subset of* α. *Then* $\beta \subseteq \delta$.

Proof. β and δ are both P-undominated subsets of α. Hence, so is $\delta \cap \beta$ unless $\delta \cap \beta = \emptyset$. But $\delta \cap \beta \neq \emptyset$, for if $\delta \cap \beta = \emptyset$, we should have $\beta \subseteq \alpha - \delta$, and thus, since δ is P-dominant, xPy for some $x \in \delta$ and $y \in \beta$, contrary to the P-undominated character of β. Consequently, $\delta \cap \beta$ is indeed a P-undominated subset of α. Thus, since β is a *minimum* P-undominated subset of α, $\delta \cap \beta \not\subset \beta$. But $\delta \cap \beta \subseteq \beta$. So $\delta \cap \beta = \beta$; that is, $\beta \subseteq \delta$. ■

BICH and GOCHA are equivalent just when P is acyclic. When are GOCHA and GETCHA equivalent?

They are equivalent when P satisfies:

P-Connexity *If $x, y \in A$ and $x \neq y$, either xPy and yPx.*

They are equivalent as well under a weaker condition:

PI-Transitivity *If $xPyIz$ then xPz.*

This follows from P-Connexity because if P-Connexity holds then no two elements of A are indifferent, so that if $xPyIz$ then $y = z$ and thus xPz. But PI-Transitivity is strictly weaker than P-Connexity because if the elements of A are all mutually indifferent and greater than two in number then PI-Transitivity holds (because xPy never holds) but P-Connexity fails.

Some facts of elementary logic merit mention at this point; I leave proofs to you. PI-Transitivity is equivalent to the conjunction of I-Transitivity and:

PP-Connexity *If $xPyPz$ then either xPz or zPx.*

This condition is what remains of P-Transitivity after one subtracts P-Acyclicity: PP-Connexity and P-Acyclicity are mutually independent and conjointly equivalent to P-Transitivity. In the same way, PI-Transitivity is what remains of P + I-Transitivity after one subtracts P-Acyclicity. These and other relationships are summarized in figure 6.2, in which an arrow (hence a chain of arrows) stands for logical

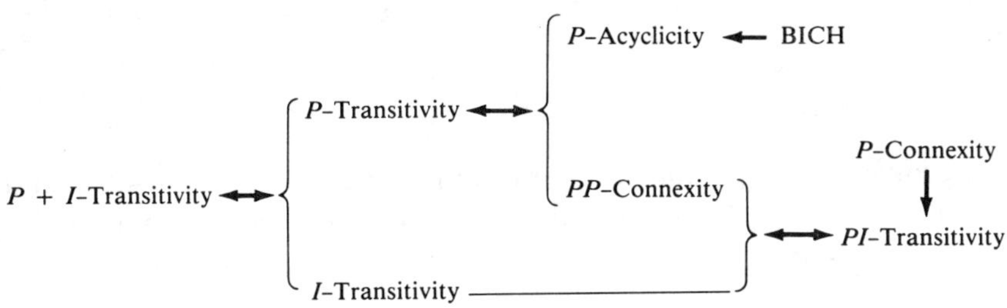

Figure 6.2

implication, the absence of an arrow (or chain) for nonimplication, and a brace for conjunction.

PI-Transitivity is not only a sufficient condition for the equality of the GOCHA and GETCHA sets; it is a necessary condition as well.

THEOREM 6.3.3. *PI-Transitivity holds iff, for every $\alpha \in S$, the minimum P-dominant subset of α is the union of minimum P-undominated subsets of α. Hence, PI-Transitivity holds iff GETCHA is equivalent to GOCHA.*

Proof. Assume PI-Transitivity and therewith *I*-Transitivity and PP-Connexity. Suppose $\alpha \in S$. Let μ be the union of minimum *P*-undominated subsets of α, and δ the minimum *P*-dominant subset of α. By Theorem 6.3.2, $\mu \subseteq \delta$. But since δ is the minimum *P*-dominant subset of α, if μ is a *P*-dominant subset of α then $\mu = \delta$. So it suffices to show that μ is a *P*-dominant subset of α. By theorem 6.1.2, μ is nonempty. Suppose $x \in \mu$ and $y \in \alpha - \mu$; to prove that xPy.

Let P^*_α be the ancestral of P/α. By Corollary 6.2.2, something in α bears P^*_α to y. Therefore, because P^*_α is transitive and α is finite, some P^*_α-undominated element z of α bears P^*_α to y. By Corollary 6.2.2 again, $z \in \mu$. So for some $r \geq 2$, there are $w_1, \ldots, w_r \in a$ such that

$$z = w_1 P w_2 P \cdots P w_r = y.$$

Because $z \in \mu$ and $y \notin \mu$, there is an $i = 1, 2, \ldots, r$ such that

$$w_i \in \mu \qquad \text{and} \qquad w_{i+1} \in \alpha - \mu.$$

For $j = i + 1, \ldots, r$, I shall show, by induction on j, that $w_i P w_j$. By construction, $w_i P w_{i+1}$. By inductive hypothesis, $w_i P w_{j-1}$. But $w_{j-1} P w_j$. So by PP-Connexity, either $w_i P w_j$ or $w_j P w_i$. Since $w_i \in \mu$ and $w_j \in \alpha - \mu$, however, not $w_j P w_i$. Hence, $w_i P w_j$, $j = 1, i + 1, \ldots, r$. In particular, then,

$$(*) \quad w_i P w_r = y.$$

Because $x, w_i \in \mu$, there are minimum *P*-undominated subsets β and γ of μ, each either the unit set of a *P*-undominated element of α or else a top *P*-cycle in α, such that

$x \in \beta$ and $w_i \in \gamma$.

Two cases:

Case 1. $\beta \cap \gamma \neq \emptyset$. Then either $\beta = \gamma = \{x\} = \{w_i\}$, or else $\beta \cup \gamma$ is a *P*-cycle. If the former, then $x = w_i$, whence xPy by (∗), and the Theorem is proved. Suppose, on the other hand, that $\beta \cup \gamma$ is a *P*-cycle. Then there are $z_1, \ldots, z_k \in \beta \cup \gamma \subseteq \mu$ such that

$$x = z_1Pz_2P \cdot \cdot \cdot Pz_k = w_i.$$

For $j = 1, 2, \ldots, k$, I shall show, by induction on $k - j$, that z_jPy. By (∗), $z_k = w_iPy$. And by inductive hypothesis, $z_{j+1}Py$. But z_jPz_{j+1}. So by PP-Connexity, either z_jPy or yPz_j. Since $z_j \in \mu$ and $y \in \alpha - \mu$, however, not yPz_j. Hence, z_jPy. Since this holds for all $j = 1, 2, \ldots, k$, we have $x = z_1Py$.

Case 2. $\beta \cap \gamma = \emptyset$. Then since β and γ are *P*-undominated subsets of α, neither xPw_i nor w_iPx. That is, xIw_i. So if yIx, then yIw_i by I-Transitivity, contrary to (∗). Hence, not yIx. Since $x \in \mu$ and $y \in \alpha - \mu$, however, not yPx. Consequently, xPy.

Now suppose PI-Transitivity does *not* hold; to deduce that for some $\alpha \in S$, the union of minimum *P*-undominated subsets of α is *not* the same as the minimum *P*-dominant subset. Then there are $x, y, z \in A$ such that either $xPyIz$ and xIz, or else $xPyIz$ and zPx. In either case, because *P* is asymmetric by definition, *x, y,* and *z* are three distinct objects. In the first case, the minimum *P*-undominated subsets of $\{x, y, z\}$ are $\{x\}$ and $\{z\}$, whereas the minimum *P*-dominant subset is $\{x, y, z\}$. In the second case, $\{z\}$ is the only minimum *P*-undominated subset of $\{x, y, z\}$, whereas $\{x, y, z\}$ itself is again the minimum *P*-dominant subset. ■

It follows immediately from Theorems 6.3.1 and 6.3.3 that the very strong P + I-Transitivity—which is equivalent to PI-Transitivity plus P-Acyclicity—is a necessary and sufficient condition for the equivalence of BICH and GETCHA.

To sum up: P + I-Transitivity is equivalent to the conjunction of two mutually independent conditions, P-Acyclicity and PI-Transitivity—the former also a consequence of BICH, the latter also a consequence of P-Connexity. BICH and GOCHA are equivalent just when P-Acyclicity holds, whereas GOCHA and GETCHA are equivalent just when PI-Transitivity holds. It follows that BICH and GETCHA are equivalent just when P + I-Transitivity holds. Regardless of transitivity conditions, the BICH set is perforce a subset of the GOCHA set, which is perforce a subset of the GETCHA set.

6.4. STABILITY PROPERTIES: AXIOMATIC CHARACTERIZATIONS OF THE GOCHA AND GETCHA FUNCTIONS

BICH implies that $C(\alpha)$ always is *internally stable*, in the sense that nothing in $C(\alpha)$ bears P to anything else in $C(\alpha)$, and *externally* stable, in the sense that nothing in $\alpha - C(\alpha)$ bears P to anything in $C(\alpha)$. GOCHA, too, guarantees *ex*ternal stability. More generally:

THEOREM 6.4.1. *Suppose x belongs to the union of minimum ρ-undominated subsets of α and y does not, although $y \in \alpha$. Then not $y\rho x$.*

Proof. By hypothesis, x belongs to some minimum ρ-undominated subset β of α whereas $y \in \alpha - \beta$. Therefore, by the ρ-undominated character of β, not $y\rho x$. ■

Now suppose $xPyPzPx$. Then $\{x, y, z\}$ is the one and only minimum P-undominated subset of itself. So $C(\{x, y, z\}) = \{x, y, z\}$ according to GOCHA. Therefore, GOCHA does *not* guarantee *in*ternal stability.

But the instabilities allowed by GOCHA are mild in a way. They are not of a sort that automatically enable us to pare down the GOCHA set. Specifically, although some members of the GOCHA set bear P to others, we can never partition this set into two subsets, one of which is clearly preferred to the other in the following sense: Some member of the first subset bears P to some member of the second, and no member of the second bears P to any member of the first. More generally:

THEOREM 6.4.2. *If U is the union of minimum ρ-undominated subsets of α, there is no subset β of U such that something in β bears ρ to something in $U - \beta$ whereas nothing in $U - \beta$ bears ρ to anything in β.*

Proof. Suppose, on the contrary, that some $x \in \beta \subseteq U$ bore ρ to some $y \in U - \beta$ whereas nothing in $U - \beta$ bore ρ to anything in β; to deduce a contradiction. Let ρ^*_α be the ancestral of ρ/α. Since nothing in $U - \beta$ bears ρ to anything in β, were there a ρ/α-chain from y to x, some term of the chain would belong to $\alpha - U$, so something in $\alpha - U$ would bear ρ to something in $\beta \subseteq U$, contrary to Theorem 6.4.1. Therefore, there is no ρ/α-chain from y to x. But (x, y) is a ρ/α-chain from x to y. Hence, $x\rho^*_\alpha y$, which is impossible by Theorem 6.2.2. because $y \in U$. ■

BICH implies, not just that $C(\alpha)$ is stable, internally and externally, not just that $C(\alpha)$ contains *only* P-undominated elements of α, but also that $C(\alpha)$ contains *all* P-undominated elements of α. More generally, BICH implies that every P-undominated *subset* of α shares a member with $C(\alpha)$. GOCHA has the same consequence.

THEOREM 6.4.3. *If α is finite and β is a ρ-undominated subset of α, some element of β belongs to the union of minimum ρ-undominated subsets of α.*

Proof. Since α is finite, some subset λ of β is a *minimum* ρ-undominated subset of α. By definition, λ must be nonempty; say $x \in \lambda$. Then x belongs to the union of minimum ρ-undominated subsets of α. ■

What you have just seen is that GOCHA shares these three consequences with BICH:

Nothing in $\alpha - C(\alpha)$ bears P to anything in $C(\alpha)$.
No subset β of $C(\alpha)$ is such that something in β bears P to something in $C(\alpha) - \beta$ and nothing in $C(\alpha) - \beta$ bears P to anything in β.
If $\alpha \in S$ and β is a P-undominated subset of α, some element of β belongs to $C(\alpha)$.

Thanks to Theorem 6.1.4(e), GETCHA has the same consequences, with P replaced by R. Because $R = A^2 - P^{-1} = P \cup I$, GETCHA's three consequences may be recast as follows:

Everything in $C(\alpha)$ bears P to everything in $\alpha - C(\alpha)$.
No subset of β of $C(\alpha)$ is such that everything in β bears P to everything in $C(\alpha) - \beta$.
If $\alpha \in S$ and β is a P-dominant subset of α, some element of β belongs to $C(\alpha)$.

Note that the first consequence of GETCHA is stronger than the first consequence of GOCHA whereas the second and third consequences of GETCHA are weaker than the second and third consequences of GOCHA.

It happens that the three consequences of GOCHA also *imply* GOCHA, so GOCHA is equivalent to their conjunction (assuming Axioms C1 and C2). The same is true of GETCHA, with P replaced by R. More generally:

THEOREM 6.4.4. *Assume Axioms C1 and C2. Suppose, for each* $\alpha \in S$, *that* ρ_α *is a binary relation on A. Then of the following four conditions, the first three are conjointly equivalent to the fourth:*

(I) *Nothing in* $\alpha - C(\alpha)$ *bears* ρ_α *to anything in* $C(\alpha)$.

(II) *No subset* β *of* $C(\alpha)$ *is such that something in* β *bears* ρ_α *to something in* $C(\alpha) - \beta$ *and nothing in* $C(\alpha) - \beta$ bears ρ_α *to anything in* β.

(III) *If* $\alpha \in S$ *and* β *is a* ρ_α*-undominated subset of* α, *some element of* β *belongs to* $C(\alpha)$.

(IV) $C(\alpha) \equiv$ *the union of minimum* ρ_α*-undominated subsets of* α.

Proof. By Theorems 6.4.1–6.4.3, (IV) implies (I)–(III).

Now assume (I)–(III). Let $\alpha \in S$ and $U =$ the union of minimum ρ_α-undominated subsets of α; to prove that $C(\alpha) = U$.

Let β be a minimum ρ_α-undominated subset of α. Then nothing in $\alpha - \beta$ bears ρ_α to anything in β, whence nothing in $C(\alpha) - \beta$ bears ρ_α to anything in $C(\alpha) \cap \beta$. And by (I), nothing in $\alpha - C(\alpha)$ bears ρ_α to anything in $C(\alpha)$, whence nothing in $\alpha - C(\alpha)$ bears ρ_α to anything in $C(\alpha) \cap \beta$. Therefore, nothing in $\alpha - (C(\alpha) \cap \beta) = (C(\alpha) - \beta) \cup (\alpha - C(\alpha))$ bears ρ_α to anything in $C(\alpha) \cap \beta$; that is, $C(\alpha) \cap \beta$ is a ρ_α-undominated subset of α if it is nonempty. Thus, since β is a *minimum* ρ_α-undominated subset of α, $C(\alpha) \cap \beta$ cannot be a nonempty proper subset of β. So either $C(\alpha) \cap \beta = \emptyset$ or $C(\alpha) \cap \beta = \beta$. But by (III), $C(\alpha) \cap \beta \neq \emptyset$. Hence, $C(\alpha) \cap \beta = \beta$; that is, $\beta \subseteq C(\alpha)$. Because our choice of β was arbitrary, every minimum ρ_α-undominated subset of α is a subset of $C(\alpha)$. So $U \subseteq C(\alpha)$, and it suffices to show that $C(\alpha) \subseteq U$.

By Theorem 6.4.1, nothing in $\alpha - U$ bears ρ_α to anything in U, whence nothing in $C(\alpha) - U$ bears ρ_α to anything in $C(\alpha) \cap U$. By (II), then, nothing in $C(\alpha) \cap U$ bears ρ_α to anything in $C(\alpha) - U$. Hence, by (I), nothing in $\alpha - (C(\alpha) - U)$ bears ρ_α to anything in $C(\alpha) - U$. Therefore, if $C(\alpha) - U$ were nonempty, it would be a ρ_α-undominated subset of α, and so by (III), some element of $C(\alpha) - U$ would belong to U, which is absurd. Consequently, $C(\alpha) - U = \emptyset$; that is, $C(\alpha) \subseteq U$. ■

Note the generality of this theorem: not only was ρ_α not assumed to be P or R, but it was allowed (although not required) to vary with α. This feature gives the theorem recurring utility in chapter 7.

If we think of GOCHA and GETCHA as definitions of C in terms of P, then Axioms C1 and C2 plus (I)–(III) with $\rho_\alpha \equiv P$ *axiomatically characterize*—are uniquely satisfied by—the GOCHA-defined choice function, and these same conditions with $\rho_\alpha \equiv R$ axiomatically characterize the GETCHA-defined function.

6.5. THE PARETO EFFICIENCY PROBLEM

A common complaint against GOCHA and GETCHA is that they tend to require $C(\alpha)$ to be too inclusive—to comprise either too many alternatives or certain specific alternatives that are not choosable. This complaint is especially likely to be leveled against GETCHA. In one of the examples offered in section 6.3, the BICH and GOCHA sets were the same unit set, yet the GETCHA set was the entire six-member feasible set.

As it stands, the "too many" objection is not too conclusive. How many are too many? Even if $C(\alpha)$ sometimes is surprisingly large, it may just be that the process represented by C is surprisingly permissive or indeterministic (section 1.2). Collective-choice assumptions with surprising consequences should by now be no surprise.

A stronger objection to GOCHA and GETCHA, from both a positive and a normative point of view, is that they require $C(\alpha)$ to contain so-called *Pareto-inefficient* alternatives—ones to which others are unanimously preferred. For constant individual preference relations $P_1, \ldots, P_n$, there are circumstances under which GOCHA and GETCHA both conflict with this condition:

PARETO EFFICIENCY *If* $x \in \alpha \in S$ *and* xP_iy *for all* $i \in N$ *then* $y \notin C(\alpha)$.

Example: $n = 3$, and the three individuals order $\alpha = \{x, y, z, w\}$ as follows:

Mr. 1	Mr. 2	Mr. 3
x	y	z
y	z	w
z	w	x
w	x	y

Since everyone prefers z to w, Pareto Efficiency implies that $w \notin C(\alpha)$. Interpreted as simple-majority preference, though, P has this form:

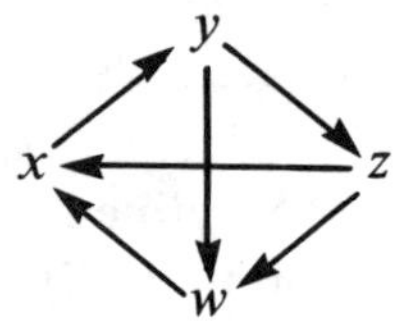

So GOCHA and GETCHA imply that $C(\alpha) = \alpha$, contrary to Pareto Efficiency.

Pareto Efficiency is an intuitively attractive assumption, Pareto inefficiencies intuitively disturbing phenomena. Why?

Although w has the trait that everyone prefers another member of α to it, *every* member of α has the trait that *a majority* prefer another member to it, and it is *majorities generally* (we may suppose) that rule. What I question is why, when P is the relation of simple-majority preference, we ought to assume Pareto Efficiency without generally assuming:

If $x \in \beta \in S$ and $|\{i \in N | xP_i y\}| > \frac{n}{2}$ then $y \notin C(\beta)$,

which is unacceptable since it requires $C(\alpha)$ in the example above to be empty. More generally, I question why we should assume Pareto Efficiency without assuming:

STABLE-CHOICE CONDITION *If* $x \in \beta \in S$ *and* xPy *then* $y \notin C(\beta)$,

which is the left-to-right half of BICH—the more controversial half, since *it* is what implies P-Acyclicity.

I discern two simple reasons theorists have had for postulating Pareto Efficiency:

Reason 1. It is reasonable to assume both

PAIRWISE PARETO EFFICIENCY *If* $xP_i y$ *for all* $i \in N$ *then* xPy

and Stable-Choice Condition. But these two assumptions obviously imply Pareto Efficiency.

Reason 2. It is reasonable to assume

INTUITIVE UNANIMITY CONDITION *If Messrs.* 1, 2, . . . , *n unanimously oppose the choice of x from* α (*if they can all agree that x should not be chosen from* α) *then* $x \notin C(a)$, and this justifies Pareto Efficiency.

The trouble with Reason 1 is the Stable-Choice Condition—the objectionable, left-to-right half of BICH. Yes, it justifies Pareto Efficiency, but at the cost of implying P-Acyclicity.

The trouble with Reason 2 is that Intuitive Unanimity Condition is not strong enough to justify Pareto Efficiency. In the last example, w is Pareto inefficient but not necessarily unanimously opposed. Whether Messrs. 1, 2, . . . , n would all agree to reject w depends on what they expect to happen as a result of rejecting w, on the intensity of their preferences, and on how they act in the face of uncertainty. If, for example, Mr. 3 disliked y sufficiently, *he* would not agree to reject w. It could well be the case that Mr. 3 would not agree to reject w unless y were rejected as well, a proposal to which Mr. 2 would not agree.

I think there *are* sound reasons to postulate Pareto Efficiency, if not for all collective-choice processes then for a large and important class. But a subtle analysis of the impact of cooperation on collective choice is needed to bring that out; I turn to this, among other things, in the next chapter.

But first I want to generalize the problem of GETCHA, GOCHA, and Pareto Efficiency. Stable-Choice Condition is equivalent to the conjunction of these two conditions:

INTERNAL NONDOMINANCE *xPy for no $x, y \in C(\alpha)$.*

EXTERNAL NONDOMINANCE *xPy for no $x \in \alpha - C(\alpha)$ and $y \in C(\alpha)$.*

Where BICH entails both these conditions, GETCHA and GOCHA just entail the second, and it is the second that can conflict with Pareto Efficiency: *any* choice function satisfying External Nondominance, not just those defined by GETCHA and GOCHA, can violate Pareto Efficiency, depending on P.

To see why, consider the last example. By Axiom C1, $C(\{x, y, z, w\}) \neq \emptyset$. Because $\{x, y, z, w\}$ is a P-cycle, if $C(\{x, y, z, w\})$ were a nonempty proper subset of $\{x, y, z, w\}$, then some element of $\{x, y, z, w\} - C(\{x, y, z, w\})$ would bear P to some element of $C(\{x, y, z, w\})$, contrary to External Nondominance. Hence, $C(\{x, y, z, w\}) \not\subset \{x, y, z, w\}$, and so, by Axiom C2, $C(\{x, y, z, w\}) = \{x, y, z, w\}$. But this contradicts Pareto Efficiency since everyone prefers z to w.

The problem is yet more general. It arises for any collective-choice process satisfying γ-Domain, Binary Independence, Weak Nonblocker, Pareto Preference, Weak Positive Responsiveness, and "$n \geq 5$." For assume these conditions, and suppose $\gamma > 4$. Then by Theorem 3.6.1, there exists a $v \in$ *PROF and a* $\mathbb{P}^v$-cycle α such that $xP^v_N y$ for some $x, y \in \alpha$. Suppose there were a choice function C satisfying External Nondominance (with $P = \mathbb{P}^v$) and Pareto Efficiency (with $P_i = P^v_i$). Then by External Nondominance, $C(\alpha)$ would have to be α itself since otherwise we should have xPy for some $x \in \alpha - C(\alpha)$ and $y \in C(\alpha)$. But that contradicts Pareto Efficiency since $xP^v_N y$ for some $x, y \in \alpha$. To put the point another way, in Theorem 3.6.1 we can replace Weak Collective P-Acyclicity by the assumption that $\gamma > 4$ plus:

> WEAK PARETIAN STABILITY *For every γ-member subset α of A and every $v \in$ PROF, there is a nonempty subset β of α such that* (1) $x\mathbb{P}^v y$ *for no* $x \in \alpha - \beta$ *and* $y \in \beta$ *and* (2) $xP^v_N y$ *for no* $x \in \alpha$ and $y \in \beta$.

Conflict with Pareto Efficiency is a problem, not just for GETCHA and GOCHA, but for any choice-set specification (or for any choice function) satisfying External Nondominance. The demand for Pareto Efficiency fairly forces us to allow situations to arise in which a feasible alternative outside the choice set is collectively preferred to one inside the choice set.

Despite the Pareto Efficiency problem, which many theorists would regard as devastating from a normative point of view, GETCHA and GOCHA are important for three reasons: First, they specify $C(\alpha)$ completely in terms of P. Second, they are derivable from ostensibly reasonable assumptions. Third and most important, whatever their normative shortcomings, they are extremely useful as positive modeling tools because the choice sets they specify are demonstrably the choice sets produced by certain conspicuous institutional procedures, as you will see in chapters 9 and 10.

CHAPTER 7

SOCO

7.1. POWER

What Messrs $1, 2, \ldots, n$ would choose collectively in a given situation depends on the underlying power structure or constitution, on who has the power to decide what. Under a purely *majoritarian* power structure—a majority game, if you prefer—every majority of individuals has the power to dictate every choice. Under a strictly *dictatorial* power structure, one man has the power to dictate every choice. Because a dictator can disagree with a majority, the collective choice under a majoritarian power structure can differ from the collective choice under a dictatorial one, although individuals and their preferences and the feasible set be the same.

Power Structures: Definition

What, in general, is a power structure?

The majoritarian and dictatorial examples suggest that a power structure is a set *POW* of groups of individuals (subsets of N), interpreted as follows:

$g \in POW$ iff g is a *universally effective group of actors*—a set of individuals with the power to dictate every collective choice.

In other words:

$g \in POW$ iff $g \subseteq N$ and, for every $\alpha \in S$ and $x \in a$, there is a combination of actions open to the members of g that would secure the choice of x from α.

This formulation is too narrow. Although groups sometimes are universally effective, the power of a group to dictate a choice often depends on the feasible set. A group *g* might have the power to dictate the choice from α but not from β. Any group comprising the president of the United States plus a majority from each house of Congress has the power to dictate a change in the U.S. Tax Code but not a Constitutional amendment.

Whether *g* has the power to dictate a given choice also might depend on the alternative whose choice *g* would dictate; *g* might have the power to dictate the choice of *x* but not of *y*. In the U.N. Security council, each permanent member, by itself, has the power to dictate *defeat* of any motion but not the power to dictate *passage* of every motion.

Whether *g* has the power to dictate a certain choice might even depend on both factors—the feasible set and the alternative whose choice *g* would dictate; *g* might have the power to dictate the choice of *x* from α but not the choice of *y* from α or of *x* from β. Example: I bequeath my pet hippopotamus to my wife, then to my daughter if my wife does not want it, then to my son if neither my wife nor my daughter wants it, then to the San Antonio Zoo if none of my family wants it. The zoo, at least, wants the hippopotamus. So either my wife will get it (*w*), or my daughter will get it (*d*), or my son will get it (*s*), or the zoo will get it (*z*). Such are the terms of my bequest that my daughter has the power to dictate the choice of *d* from $\{d, s, z\}$ but not the choice of *s* from $\{d, s, z\}$ nor the choice of *d* from $\{w, d, s, z\}$.

This suggests that a power structure is a ternary relation *POW*, interpreted thus:

> $POW(g, x, \alpha)$ iff $g \subseteq N$, $x \in \alpha \in S$, and *g* has the power to dictate the choice of *x* from α.

Even this is too narrow. A group's power to dictate a choice might depend on which alternative choice would be blocked thereby—on what the choice would be without the group's intervention. Maybe *g*'s *x*-strategy—what *g* must do to secure the choice of *x*—is not generally effective but is effective as a response to some *y*-strategy. So if *y* would be chosen without *g*'s intervention then *g* can dictate the choice of *x* instead; but if it were *z* rather than *y* that would be chosen without *g*'s intervention then perhaps *g* would be powerless to dictate the choice of *x*. In the hippopotamus-bequest example, my son can dictate the choice of s from $\{w, d, s, z\}$ to block the choice of z from $\{w, d, s, z\}$ but not to block any other choice from $\{w, d, s, z\}$: by demanding s, he can block z, but he cannot thereby secure s unless w and z have been ruled out; only if the choice would otherwise be z can his intervention secure s.

What all this suggests is that a power structure may be represented by a quaternary relation *POW*, interpreted as follows:

> $POW(g, x, y, \alpha)$ iff $g \subseteq N$, $x, y \in \alpha \in S$, and g has the power to block the choice of y from α in favor of x. That is, for every y-strategy, there is an effective counterstrategy open to g, the result of which would be x.

To put a bit of flesh on these bones, suppose $g \subseteq N$ and Messrs. 1, 2, . . . , n are to make a collective choice from a set α containing at least two alternatives, x and y. This choice will be determined by a *scenario*—a combination of actions by Messrs. 1, 2, . . . , n. Then $POW(g, x, y, \alpha)$ holds if, for every y-scenario—every scenario that would result in y—there is an x-scenario that differs from this y-scenario only in some combination of actions by some members of g that is feasible for g in the presence of the given y-scenario. This combination of actions includes anything the g's may have to do to plan and coordinate their behavior and secure one another's cooperation.

Here is a more concise interpretation:

> $POW(g, x, y, \alpha)$ iff $g \subseteq N$, $x, y \in \alpha \in S$, and g has the power to move from y to x in α.

This provides a convenient verbal paraphrase of statements about *POW* relationships.

One might object even to this way of representing power structures on the ground that g's power to block the choice of y from α in favor of x could conceivably depend on *which of several* y-strategies g is responding to. Maybe g has the power to turn *some but not all* y-scenarios into x-scenarios.

This problem is easily avoided by drawing fine enough distinctions among collective-choice alternatives—by construing otherwise-similar alternatives as distinct if they are determined by distinct scenarios, at least in cases of the troublesome sort just described. (This could force us to treat the choice of the same political candidate by two distinct vote combinations as two distinct, although importantly similar, alternatives.)

I propose identifying power structures with quaternary relations interpreted as above.

Four clarifications:

First, this way of representing power structures *allows* us to say that the power to do certain things depends on certain factors, but it does *not require* that this power depend on those factors. For example, it allows us to say that a group's power to dictate a choice depends on the alternative choice blocked thereby, but it also allows us to say that this power does *not* depend on the alternative choice blocked thereby—to say that if $POW(g, x, y, \alpha)$ and $x \neq y$ then $POW(g, x, z, \alpha)$ for all $z \neq x$ in α.

Second, the ability of a group to dictate a choice does *not* require that its members *coalesce*—that they *act cooperatively*. They need not communicate, agree on a joint strategy, trust one another, or be trustworthy. For $POW(g, x, y, \alpha)$ to hold, it is enough that any y-scenario can be turned into an x-scenario by some combination of actions that is open to g in the presence of the given y-scenario. This combination of actions might require cooperation. Then again, it might not. Example: A legislature is to vote on two bills, each supported only by a minority. If each legislator votes against the bill he opposes then the defeat of each bill will have been dictated by some majority without any cooperation. But the supporters of the two bills, taken together, constitute a majority, each of whom prefers passage of both bills to defeat of both. If they trade votes, each agreeing to vote for the bill he opposes, then this combined group will dictate passage of both bills, but only by cooperating—by communicating, agreeing on a joint course of action, and being faithful.

Third, one collective-choice situation can contain any number of power structures, differing in the *kinds of power* they represent. In the last example, the "majority of minorities" has the *de jure* power to dictate the choice of both bills: the legislative rules explicitly give every majority this power. In another sense, the same group lacks this power if its members cannot cooperate—if they cannot afford the cost of communicating, enforcing their agreement, and so on.

Fourth, my interpretation of *POW* assumes a fixed profile of individual preferences ($P_1, \ldots, P_n$). I have therefore allowed the possibility that power structures depend on, and vary with, individual preferences.

Properties of Power Structures

To represent a power structure, a quaternary relation *POW* must fulfill four axioms:

AXIOM *POW*1 *If* $POW(g, x, y, \alpha)$ *then* $g \subseteq N$, $g \neq \emptyset$, *and* $x, y \in \alpha \in S$.

This just says that *POW* always relates a nonempty group of individuals to a pair of alternatives and a potential feasible set containing them.

AXIOM *POW*2 *If* $POW(g, x, y, \alpha)$ *and* $POW(g, y, z, \alpha)$, *then* $POW(g, x, z, \alpha)$.

According to this axiom, power is *transitive:* if g has the power to move from y to x and from z to y in α then g has the power to move from z to x in α—g can do this by moving from z to y and from there to x.

AXIOM *POW* 3 *If* $POW(g, x, y, \alpha)$ *and* $g \subseteq h \subseteq N$ *then* $POW(h, x, y, \alpha)$.

This says power is *monotonic*, or *superadditive:* any power possessed by any group is possessed as well by any more inclusive group. For if $g \subseteq h \subseteq N$ and g has the power to move from y to x in α then h can move from y to x by a combination of actions consisting in g's moving from y to x and $(h - g)$'s doing nothing.

AXIOM *POW*4 *If* xP_iy *for every* $i \in g$ *and* yP_ix *for every* $i \in h$ *then not both* $POW(g, x, y, \{x, y\})$ *and* $POW(h, y, x, \{x, y\})$.

This says the assignment of power to dictate pairwise choices is *consistent*.

These axioms have a useful consequence:

THEOREM 7.1.1. *Assume Axioms POW*2 *and* 3. *Suppose* $POW(g_1, x, y, \alpha)$, $POW(g_1, y, z, \alpha)$, *and* $g_1 \cup g_2 \subseteq g_3 \subseteq N$. *Then* $POW(g_3, x, z, \alpha)$.

Proof. $POW(g_3, x, y, \alpha)$ and $POW(g_3, y, z, \alpha)$ by Axiom *POW* 3, whence $POW(g_3, x, z, \alpha)$ by Axiom *POW* 2. ■

Group power, as I have represented it, is a function of two

alternatives and a feasible set. Depending on context, it might be independent of any or all of these factors. Under some power structures, power is independent of one or both alternatives:

If $POW(g, x, y, \alpha)$ and $x \neq y$ then $POW(g, x, z, \alpha)$ for every $z \in \alpha - \{x\}$.

If $POW(g, x, y, \alpha)$ and $x \neq y$ then $POW(g, z, w, \alpha)$ for all distinct $z, w \in \alpha$.

And under many power structures, power is independent of the feasible set—a condition I call:

BINARY POWER *If* $POW(g, x, y, \alpha)$ *and* $x, y \in \beta \in S$ *then* $POW(g, x, y, \beta)$.

This says that if a group ever has the power to move from y to x then it has the power to move from y to x in every potential feasible set containing x and y. Binary Power will play an important role in section 7.4.

Another frequently satisfied condition that will play an important role in section 7.4 is:

UNRESTRICTED ACCESS *If* $x, y \in \alpha \in S$ *and* $x \neq y$ *then* $POW(g, x, y, \alpha)$ for some g.

This says that for any two feasible alternatives x and y, *some* group has the power to move from y to x.

The meaning of Unrestricted Access is brought out most clearly by one of its consequences: Unrestricted Access plus Axioms *POW* 1 and 3 imply that Messrs. $1, 2, \ldots, n$ can accomplish anything when they act unanimously.

THEOREM 7.1.2. *Assume Unrestricted Access and Axioms POW* 1 and 3. Then $POW(N, x, y, \alpha)$ *for all* $\alpha \in S$ *and all distinct* $x, y \in \alpha$.

Proof. $POW(g, x, y, \alpha)$ for some $g \subseteq N$ by Axiom *POW* 1 and Unrestricted Access, whence $POW(N, x, y, \alpha)$ by Axiom *POW* 3. ■

7.2. EXERCISE

Power is not always exercised. In 1936 each of many sets of voters had the power to elect Alfred M. Landon president. None exercised that power. Collective choices depend not only on power but on the incentive—the rational willingness—to exercise it.

Let E be the quaternary relation interpreted this way:

$E(g, x, y, \alpha)$ iff $g \subseteq N$, $x, y \in \alpha \in S$, and g has both the power *and the incentive* to move from y to x in α—to dictate the choice of x from α in order to block the choice of y from α.

This is just a rough interpretation of E; it admits of alternative refinements. How precisely to define E?

The obvious suggestion is to equate E with E_1, defined thus:

DEFINITION $E_1(g, x, y, \alpha)$ *iff* $POW(g, x, y, \alpha)$ *and* xP_iy *for all* $i \in g$.

On this view, a group with the power to move from y to x in α has an *incentive* to do so just in case its members *all prefer* x to y.

There is a problem with this approach. Suppose $E_1(g, x, y, \alpha)$ and $E_1(h, z, x, \alpha)$. Then, one might argue, g does not have *sufficient incentive* to move from y to x in α if $g \subseteq h$, and g does not have *sufficient power* to move from y to x in α if $g \cap h = \emptyset$—if g and h share no members. For if $g \subseteq h$, then x is not the best g can do: because $E_1(h, z, x, \alpha)$, the g's would all be better off supporting z (in company with the members, if any, of $h - g$) than supporting x. And if $g \cap h \neq \emptyset$ then g cannot secure x: because $E_1(h, z, x, \alpha)$, some actors foreign to g—the members of h—are able and willing to block x in favor of z. The point is that *even if* the g's can stick together (function as a coalition, prevent defection), they still will be unwilling to support x if $g \subseteq h$, and unable to do so if $g \cap h = \emptyset$.

To avoid this objection, we might define E so that $E(g, x, y, \alpha)$ holds only if the following two conditions are satisfied:

(1) $E_1(g, x, y, \alpha)$.

(2) There do not exist h, z such that $E_1(h, z, x, \alpha)$ and either $g \subseteq h$ or $g \cap h = \emptyset$.

But this is too strong. Suppose $E_1(g, z_1, x, \alpha)$, $E_1(g, z_2, z_1, \alpha)$, $E_1(g, z_3, z_2, \alpha)$, and so on without end, because the members of g share a cyclic preference. Then (2) is violated in the extreme: x is not the best g can do, nor is z_1, nor z_2, nor anything else. Yet it is going a bit far to deny that g has sufficient power and incentive to move from y to x. True, x is not the best g can do. But nothing is. Still here must be some stopping point, however arbitrary.

Or suppose $h \cap g = \emptyset$ and $E_1(h, z_1, x, \alpha)$, $E_1(g, z_2, z_1, \alpha)$, $E_1(h, z_3, z_2, \alpha)$, and so on without end. Again (2) is violated in the extreme: g cannot secure x because h, which shares no member with g, is able and willing to move from x to z_1, whence g is able and willing to move to z_2, and so on and on. But again it is going a bit far to deny that g has sufficient power and incentive to move from y to x. Although g cannot secure x, no possible stopping point in the chain of moves away from x can be secured. Yet these moves must eventually stop, however arbitrarily.

What we need to do is weaken (2) so that it does not apply in the face of an infinite series of moves like those just sketched, lest we imply that *no* movement can take place in such a case. This is accomplished if we replace (2) with:

(2′) $SUF(g, x, y, \alpha)$,

where:

DEFINITION *$SUF(g, x, y, \alpha)$ iff for no k do there exist $z_0, z_1, \ldots, z_k$, $h_1, \ldots, h_k$ such that*

(a) $z_0 = y \neq x = z_1$;

(b) $h_1 = g$;

(c) $E_1(h_i, z_i, z_{i-1}, \alpha)$, $i = 1, 2, \ldots, k$;

(d) $h_{i-1} \subseteq h_i$ *or* $h_{i-1} \cap h_i = \emptyset$; $i = 2, 3, \ldots, k$; *and*

(e) *there do not exist h, z for which $E_1(h, z, z_k, \alpha)$ and either $h_k \subseteq h$ or $h_k \cap h = \emptyset$.*

So $SUF(g, x, y, \alpha)$ holds if, were g to move from y to x in α, then g would have *sufficient* power and incentive to stop at x: the move could be the last if g sticks together.

But even this approach goes too far. Let $\alpha = \{x, y, z\}$, and suppose $E_1(\{1\}, y, z, \alpha)$, $E_1(\{1, 2\}, x, y, \alpha)$, and there are no other E_1 relationships within α. Then not $SUF(\{1\}, y, z, \alpha)$: y is not the best $\{1\}$ can do. So according to the suggested definition of E, not $E(\{1\}, y, z, \alpha)$. This is unacceptable. If not $E(\{1\}, y, z, \alpha)$ then *no* group has the power and incentive to block z. Surely, though, $\{1\}$ has the power and incentive to block z—to move away from z. Because y is not the best $\{1\}$ can do, $\{1\}$ does not have sufficient incentive to *stop for good* at y. But $\{1\}$ does have sufficient power and incentive to block z, by moving to y on the way to x.

This shows that we should require (2′), not at every move a group makes, but only at the *last in any chain* of moves. Therefore, we might reasonably equate E with E_2, defined as follows:

DEFINITION $E_2(g, x, y, \alpha)$ *iff for some m there are* $x_0, x_1, \ldots, x_m, g_1, \ldots, g_m$ *such that*
(a) $x_0 = y \neq x = x_m$;
(b) $g_k = g$;
(c) $E_1(g_i, x_i, x_{i-1}, \alpha)$, $i = 1, 2, \ldots, m$; *and*
(d) $SUF(g, x, x_{m-1}, \alpha)$.

Note that $E_2(g, x, y, \alpha)$ can hold even if g does not *directly* have any power or incentive to move from y to x in α, so long as (1) g has the power and incentive in the E_1 sense to make the *last in a chain* of E_1 moves beginning with y and ending with x in α and (2) g has sufficient power and incentive to stop at x: g's move to x could be the last if g sticks together.

Depending on context, it might be desirable to interpret E so that it is a proper subrelation of E_2. For it might be possible in some contexts to say more about coalition behavior than simply that members achieve mutual advantage. Example: $E_2(g, x, y, \alpha)$ and $E_2(h, z, y, \alpha)$, and h is more likely than g to coalesce—because it is smaller or more homogeneous or better organized than g, or because its ideal points in some issue space (see section 3.1) are closer together, or whatever. Then we might reasonably deny that $E(g, x, y, \alpha)$, interpreting E not as E_2 but in some stronger way. Another example: $E_2(g, x, y, \alpha)$ and $E_2(g, z, y, \alpha)$, and g is more likely to agree to z than to agree to x according to some reasonable theory of bargaining. Then we might again deny that $E(g, x, y, \alpha)$.

Special contexts also might allow us to define E so that it embodies *strategic incentives.* If $POW(g, x, y, \alpha)$ but the g's do not prefer x to y, we might still say they have an incentive to move from y to x if other groups have the power and incentive to make a series of moves away from x leading to an ultimate outcome which the g's prefer to x and y.

Some of the things I shall say about the exercise of power are most easily formulated in terms of the functions D_1 and D_2, defined as follows:

DEFINITION *For* $i = 1, 2$, D_i *is the function on S whose value,* D_i^α, *at each* $\alpha \in S$ *is the relation* $\{(x, y) | E_i(g, x, y, \alpha)$ *for some* $g\}$.

So $xD_i^\alpha y$ holds whenever $E_i(g, x, y, \alpha)$ holds for some g or other.

7.3. SOLUTIONS

Collective choices are determined by participants' preferences and the power structure in which they operate—the rules of the game they play. Given these factors, what can we say in general about collective choices? That depends on how we interpret the notion of *power plus incentive*—on whether we define E as E_1 or as E_2 or in some more specialized way. But whichever way we interpret E_i, these three conditions seem reasonable:

EXTERNAL STABILITY *Nothing in* $\alpha - C(\alpha)$ *bears* D_i^α *to anything in* $C(\alpha)$.

This says that if an alternative x is choosable from α (if $x \in C(\alpha)$) then no group has the power and incentive to block x in favor of something else that could not itself have been chosen: no group has the power and incentive to move from x to y unless y is choosable as well (not $yD_i^\alpha x$ unless $y \in C(\alpha)$).

WEAK INTERNAL STABILITY *No subset* β *of* $C(\alpha)$ *is such that something in* β *bears* D_i^α *to something in* $C(\alpha) - \beta$ *but nothing in* $C(\alpha) - \beta$ *bears* D_i^α *to anything in* β.

This says the choice set can never be partitioned into two subsets in such a way that some group has the power and incentive to move from one subset (β) to the other ($C(\alpha) - \beta$) but no group has the power and incentive to move in the reverse direction. The effect of this axiom is to keep $C(\alpha)$ from being overly inclusive. I call the condition *Weak* Internal Stability because it allows members of $C(\alpha)$ to bear D_i^α to one another, so long as such relationships do not hold in just one direction across some line drawn through $C(\alpha)$.

STABLE-CHOICE INCLUSION *If* $\alpha \in S$ *and* β *is a* D_i^α*-undominated subset of* α, *some element of* β *belongs to* $C(\alpha)$.

This says that if no group has the power and incentive to move out of a nonempty set β of feasible alternatives, there is no need to move out of β in making a collective choice: at least one member of β belongs to the choice set. Here is the intuitive rationale: Let $\beta = \{x_1, \ldots, x_k\}$. Even if every proposal of the form

Let us choose x_i!

is opposed by some group with the power to prevent its adoption, no group has the power plus incentive to block adoption of the weaker proposal:

Let us choose either x_1 or x_2 or . . . or x_k!

So this proposal might be adopted, which means that either x_1 or x_2 or . . . or x_k might be chosen, which in turn means that $C(\alpha)$ must contain at least one of these alternatives.

These three conditions are conjointly equivalent to

SOLUTION CONDITION *i* (SOCO *i*) $C(\alpha) \equiv$ *the union of minimum* D_i^α-*undominated subsets of* α.

And this has further interesting equivalents.

THEOREM 7.3.1 *The following four conditions are equivalent:*

SOCO i

The conjunction of External Stability, Weak Internal Stability, and Stable-Choice Inclusion

$C(\alpha) \equiv$ *the union of all unit sets of* D_i^α-*undominated elements of* α *and all top* D_i^α-*cycles in* α

$C(\alpha) \equiv \{x \in a \mid x$ *is undominated under the ancestral of* $D_i^\alpha\}$

Proof. Immediate from Theorems 6.2.1 and 6.2.2. ■

SOCO *i* is not really a single condition but a *schema*. Its instances are conditions involving different interpretations of E_i. I am mainly concerned with SOCO 1, in which $E_i = E_1$, and SOCO 2, in which $E_i = E_2$. But further interpretations of E_i(for which I sketched some suggestions at the end of the previous section) yield further instances of SOCO *i*.

Thanks to special features of the definition of E_2, SOCO 2 has a simpler equivalent:

SOCO 2* $C(\alpha) \equiv \{x \in \alpha \mid y\overrightarrow{D}_2^\alpha x$ *for no* $y \in \alpha\}$,

where:

DEFINITION $\overrightarrow{D_2}$ *is the function on S whose value,* $\overrightarrow{D}_2^\alpha$, *at each* $\alpha \in S$ *is the relation* $\{(x, y)\} \mid xD_2^\alpha y$ *but not* $yD_2^\alpha x\}$.

LEMMA 7.3.1. *Suppose* $\alpha \in S$. *Then* $xD_2^\alpha y$ *iff there is a* D_2^α*-chain from x to y.*

Proof. If $xD_2^\alpha y$ then (x, y) is a D_2^α-chain from x to y. Suppose, on the other hand, that there is a D_2^α-chain from x to y. But by the definition of E_2, whenever $zD_2^\alpha w$ then there is a D_1^α-chain $(z_1, \ldots, z_k)$ from $z = z_1$ to $w = z_k$ such that $E_1(g, z_1, z_2, \alpha)$ and $SUF(g, z_1, z_2, \alpha)$ for some g. So there is a D_1^α-chain $(x_1, \ldots, x_m)$ from $x = x_1$ to $y = x_m$ such that $E_1(g, x_1, x_2, \alpha)$ and $SUF(g, x_1, x_2, \alpha)$ for some g, whence it follows by the definition of E_2 that $E_2(g, x_1, x_m, \alpha)$. So $x_1 = xD_2^\alpha y = x_m$. ■

THEOREM 7.3.2. *SOCO 2 is equivalent to SOCO 2*.*

Proof. For $\alpha \in S$ let $\hat{D}_2^\alpha$ be the ancestral of D_2^α. By Theorem 7.3.1, it suffices to prove:

$y\hat{D}_2^\alpha x$ iff $y\vec{D}_2^\alpha x$.

But if $y\hat{D}_2^\alpha x$ then there is a D_2^α-chain from y to x and none from x to y, whence $yD_2^\alpha x$ but not $xD_2^\alpha y$ by Lemma 7.3.1; that is, $y\vec{D}_2^\alpha x$. Conversely, if $y\vec{D}_2^\alpha x$ then $yD_2^\alpha x$ and not $xD_2^\alpha y$, whence by Lemma 7.3.1 there is a D_2^α-chain from y to x and none from x to y; that is, $y\hat{D}_2^\alpha x$. ■

7.4. BICH, GETCHA, GOCHA, AND SOCO

Just how are BICH, GETCHA, GOCHA, and the SOCO conditions related? Let us begin with SOCO 1 and SOCO 2. In the first place, these two conditions agree for all pair sets.

THEOREM 7.4.1. *Let i be 1 or 2. Assume SOCO i and Axiom POW 4. Then xPy iff* $xD_1^{\{x,y\}}y$.

Proof. By definition of E_1 and E_2, $D_1^\alpha = D_2^\alpha$ when α is a pair set belonging to S. So it suffices to prove that xPy iff $xD_i^{\{x,y\}}y$.

Suppose xPy; that is, suppose $x, y \in A$, $x \neq y$, and $C(\{x, y\}) = \{x\}$. Then by SOCO i, $\{y\}$ is not a $D_i^{\{x,y\}}$-undominated subset of $\{x, y\}$, and thus $xD_i^{\{x,y\}}y$.

Conversely, suppose $xD_i^{\{x,y\}}y$, so that $xD_1^{\{x,y\}}y$. Then by Axiom *POW* 4 and the definition of E_1, not $yD_1^{\{x,y\}}x$. Therefore, $\{x\}$ is a minimum $D_1^{\{x,y\}}$-undominated subset of $\{x, y\}$, and $\{y\}$ is not. By SOCO i, then, $C(\{x, y\}) = \{x\}$. And since $xD_1^{\{x,y\}}y$ and not $yD_1^{\{x,y\}}y$, we have $x, y \in A$ and $x \neq y$. Hence, xPy. ■

SOCO 1 and SOCO 2 do not necessarily agree for sets of more than two alternatives. Suppose $\alpha = \{x, y, z\}$, $E_1(\{1, 2\}, x, y, \alpha)$, $E_1(\{1\}, y, z, \alpha)$, and $E_1(\{2\}, z, x, \alpha)$. These E_1 relationships may be depicted as follows:

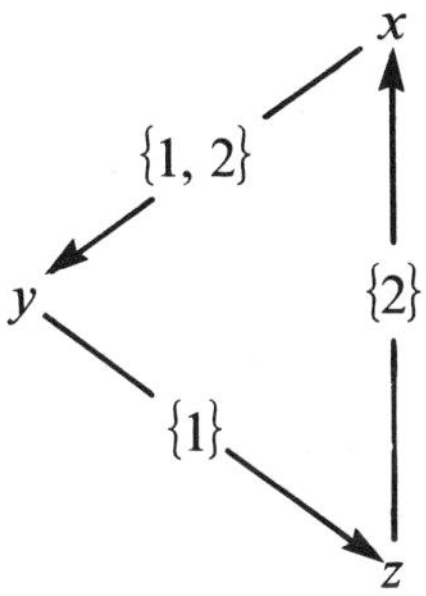

According to SOCO 1, $C(\alpha) = \alpha$. Because $SUF(\{1, 2\}, x, y, \alpha)$ but neither $SUF(\{1\}, y, z, \alpha)$ nor $SUF(\{2\}, z, x, \alpha)$ (y is not the best $\{1\}$ can do, and $\{2\}$ cannot secure z), E_2 is given by all nonreflexive chains of arrows that begin with x—{1, 2)}→y. So x bears D^α_2 to y and z, but neither of these bears D^α_2 to x. Consequently, SOCO 2 implies that $C(\alpha) = \{x\}$.

Although the set specified by SOCO 1 and the set specified by SOCO 2 are not always the same, the SOCO 2 set is always a subset of the SOCO 1 set.

THEOREM 7.4.2. *Suppose $\alpha \in S$ and x belongs to the union of minimum D^α_2-undominated subsets of α. Then x belongs to the union of minimum D^α_1-undominated subsets of α.*

Proof. Let $\hat{D}^\alpha_1$ be the ancestral of D^α_1. By Theorems 6.2.2 and 7.3.2, it suffices to prove that if something bears $\hat{D}^\alpha_1$ to x then something bears $\overrightarrow{D}^\alpha_2$ to x.

By hypothesis, there is a D^α_1-chain $(x_0, x_1, \ldots, x_m)$ from x_0 to $x_m = x$ and no D^α_1-chain from $x = x_m$ to x_0. Hence, there are $g_1, \ldots, g_m$ such that

(1) $E_1(g_i, x_{i-1}, x_i, \alpha), \qquad i = 1, 2, \ldots, m.$

Two cases:

Case 1. $SUF(g_1, x_0, x_1, \alpha)$. Then $E_2(g_1, x_0, x_1, \alpha)$, whence $x_0 D^\alpha_2 x_m = x$. And by definition of E_2, if we had $x_m = x D^\alpha_2 x_0$, there would be a D^α_1-

chain from $x = x_m$ to x_0, which there is not. So we do not have $xD_2^\alpha x_0$. Hence, $x_0 \overrightarrow{D}_2^\alpha x$.

Case 2. Not $SUF(g_1, x_0, x_1, \alpha)$. Then by definition of SUF, there are $z_0, z_1, \ldots, z_k, h_1, \ldots, h_k$ such that

(2) $z_0 = x_1 \neq x_0 = z_1$;
(3) $h_1 = g_1$;
(4) $E_1(h_i, z_i, z_{i-1}, \alpha)$, $\quad i = 1, 2, \ldots, k$; and
(5) there do not exist h, z for which $E_1(h, z, z_k, \alpha)$ and either $h_k \subseteq h$ or $h_k \cap h = \emptyset$.

By virtue of (5), $SUF(h_k, z_k, z_{k-1}, \alpha)$, whence by (1), $E_2(h_k, z_k, x_m, \alpha)$, and thus

$z_k D_2^\alpha x_m = x.$

As before, if $xD_2^\alpha z_k$, there would be a D_1^α-chain from x to z_k, and thus, by (2) and (4), there would be a D_1^α-chain from x to x_0, which there is not. So not $xD_2^\alpha z_k$. Hence, $z_k \overrightarrow{D}_2^\alpha x$. ■

What about the relation of the SOCO sets to the BICH, GOCHA, and GETCHA sets?

In answering this question, we may fairly assume:

DP-Condition $xD_1^{\{x,y\}}y$ *iff* xPy.

By Theorem 7.4.1, this follows from SOCO 1 and from SOCO 2. And given my interpretation of *POW*, *DP*-Condition is necessary truth anyway.

The two SOCO sets can simultaneously be disjoint from the BICH, GOCHA, and GETCHA sets. Let i by 1 or 2. Suppose I have total power to dictate the choice from every *three*-element set, and I prefer x to both y and z. Then

$xD_i^{\{x,y,z\}}y$ and $xD_i^{\{x,y,z\}}z.$

Suppose you have total power to dictate the choice from every *pair* set, and you prefer y to both x and z. Then

$yD_i^{\{x,y\}}x$ and $yD_i^{\{y,z\}}z,$

whence, by *DP*-Condition,

yPx and $yPz.$

According to BICH, GETCHA, and GOCHA,

$C(\{x, y, z\}) = \{y\}$.

But according to SOCO *i*,

$C(\{x, y, z\}) = \{x\}$.

In the example, power was not binary: whether a group had the power to move from one alternative to another depended on the feasible set. When *POW* satisfies Binary Power, however, the BICH set is always a subset of the two SOCO sets, the SOCO 1 set is always identical to the GOCHA set (hence is always included in the GETCHA set), and the SOCO 2 set is always a subset (possibly a proper subset) of the GOCHA set and therewith the GETCHA set.

THEOREM 7.4.3. *Assume DP*-Condition and Binary Power. Let i be 1 or 2. Then if $\alpha \in S$ *and x is a P-undominated element of* α, *x belongs to the union of minimum* D_i^α*-undominated subsets of* α.

Proof. By Theorem 7.3.1 it suffices to prove that nothing bears D_i^α to x. By the definition of E_2, if something bore D_i^α to x then, even if $i = 2$, some z would bear D_1^α to x, whence it follows by Binary Power that $zD_1^{\{x,y\}}x$, and thus, by *DP*-Condition, zPx, contrary to the hypothesis of the Theorem. ■

THEOREM 7.4.4. *Assume DP*-Condition and Binary Power. Then if $\alpha \in S$, *the union of minimum P-undominated subsets of* α *is identical to the union of minimum* D_1^α*-undominated subsets of* α. *Hence, GOCHA is equivalent to SOCO 1.*

Proof. It follows from Binary Power that for all $x, y \in \alpha$,

$xD_1^\alpha y$ iff $xD_1^{\{x,y\}}y$,

whence by *DP*-Condition,

$xD_1^\alpha y$ iff xPy.

So $P/\alpha = D_1^\alpha$. From this the Theorem follows immediately. ■

THEOREM 7.4.5. *Assume DP*-Condition and Binary Power. Then if $\alpha \in S$, *the union of minimum* D_2^α*-undominated subsets of* α *is a subset of the union of minimum P-undominated subsets of* α.

Proof. Immediate from Theorems 7.4.2 and 7.4.4. ■

This Theorem does not hold in reverse: the GOCHA set is not necessarily a subset of the SOCO 2 set, even assuming *DP*-Condition and Binary Power. Earlier I gave an example to prove that the SOCO 1 set, which is identical to the GOCHA set when Binary Power holds, is not necessarily a subset of the SOCO 2 set. And that example is consistent with *DP*-Condition plus Binary Power.

As you saw in section 6.5, the GOCHA and GETCHA sets can contain *Pareto-inefficient* (unanimously dispreferred) alternatives, even when P is the majority-preference relation. Owing to Theorem 7.4.4, the same infirmity can afflict the SOCO 1 set. By contrast, the SOCO 2 set must be Pareto efficient so long as $P_1, \ldots, P_n$ are not all cyclic and *POW* satisfies Unrestricted Access and Axioms *POW* 1 and 3.

THEOREM 7.4.6 *Assume Axioms POW* 1 and 3, Unrestricted Access, and SOCO 2. Suppose $x, y \in \alpha \in S$, *xP_iy for every $i \in N$, and P_i/α is acyclic for some $i \in N$. Then $y \notin C(\alpha)$.*

Proof. Thanks to Theorem 7.1.2, it suffices to show that $x\overrightarrow{D}{}^{\alpha}_{2}y$. But by Theorem 7.3.2, $POW(N, x, y, \alpha)$. Thus, since xP_iy for all $i \in N$,

(1) $E_1(N, x, y, \alpha)$.

By definition of *SUF*, if $SUF(N, x, y, \alpha)$ fails to hold or if $SUF(g, y, z, \alpha)$ holds for some g and z, then there is an infinite sequence $(z_1, z_2, \ldots)$ such that

$$E_1(N, z_1, x, \alpha) \text{ and } E_1(N, z_{i+1}, z_i, \alpha) \text{ for all } i,$$

in which case, because α is finite, we have $z_k = z_j$ for some j and some $k > j$, so

$$z_kP_iz_{k-1}P_iz_{k-2} \cdots z_{j+1}P_iz_j = z_k$$

for every $i \in N$, contrary to our acyclicity assumption. Hence,

(2) $SUF(N, x, y, \alpha)$ and
(3) $SUF(g, y, z, \alpha)$ for no g, z.

But by (1) and (2) we have $E_2(N, x, y, \alpha)$ and thus

$$xD^{\alpha}_{2}y.$$

And by (3), $E_2(g, y, z, \alpha)$ for no g, z, whence y bears D^{α}_{2} to nothing. Consequently, $x\overrightarrow{D}{}^{\alpha}_{2}y$. ■

The acyclicity assumption is essential to his theorem. Suppose $\alpha = \{x, y, z\}$, $POW(N, w, v, \alpha)$ for all distinct $w, v \in \alpha$, $xP_1yP_1zP_1x$, and $N = \{1\}$. Then $E_1(N, x, y, \alpha)$, $E_1(N, y, z, \alpha)$, and $E_1(N, z, x, \alpha)$. Because $n = 1$ and Mr. 1 has a cyclic preference, all three alternatives in α are Pareto inefficient. Yet SOCO 2 implies that $C(\alpha) = \alpha$. To exclude the Pareto-inefficient alternatives from $C(\alpha)$ would be to exclude everything from $C(\alpha)$.

Theorem 7.4.6 provides the rationale for Pareto efficiency that eluded us in section 6.5. Once extreme cases are contemplated, it is obvious that we need some sort of mild acyclicity assumption if we are to justify excluding Pareto-inefficient choices. And we of course need something like Axioms *POW* 1 and 3 and Unrestricted Access—or in place of these we might simply assume that $POW(N, x, y, \alpha)$ whenever $x, y \in \alpha \in S$ and $x \neq y$. The missing piece was SOCO 2.

CHAPTER 8

CHOOSING BY VOTING

8.1. THE PROBLEM

The oldest problem of collective-choice theory and still one of the liveliest is that of finding a *reasonable voting rule.* From one point of view—mine—the problem is of a piece with that of finding a reasonable choice-set or solution concept for collective-choice processes. But I had rather not say why just yet.

As the problem is customarily conceived, to solve it one must devise a reasonable voting rule with these four features:

Feature 1. It can be represented by a choice function satisfying Axioms C1 and C2 and interpreted as in sections 1.1 and 1.2, hence applicable to every finite range of mutually exclusive alternatives of some fixed but arbitrary kind.

Feature 2. Voters vote by *ranking* the feasible alternatives in order of preference or by specifying part of the information contained in such a ranking—by specifying one top-ranked alternative, or the top-ranked and second-ranked alternatives, or some such thing. A *ranking* is at least a *weak ordering.* Sometimes it is assumed, more specifically, to be a *linear* ordering.

Feature 3. The rule reflects the way voters vote and no other information.

Feature 4. Its restriction to two-alternative feasible sets is just Simple Majority Rule.

Regarding Feature 2, when I criticize various voting rules I shall only cite examples in which voters rank the feasible alternatives linearly, inasmuch as linearity is the toughest ordering condition one might impose on voter preferences. But when I propose a rule of my own, I shall not require voter preferences to be linear, transitive, or even acyclic.

Feature 3 is hardly a Universal Axiom of Reasonableness. But a rule that lacked this feature would not be a pure *voting* rule: it would reflect some information not obtainable by taking a vote. If we simply dropped Feature 3, without otherwise constraining the rule-relevant information, the problem would be too ill defined to be worth discussing.

According to Feature 4, if C represents a rule of the sort we are concerned with, then

xPy iff xMy,

where:

DEFINITION xMy *iff* $|\{i \mid xP_iy\}| > |\{i \mid yP_ix\}|$.

So the problem can be described as that of generalizing Simple Majority Rule from the two-alternative to the multialternative case.

This problem, by the way, was virtually the sole preoccupation of collective-choice theorists before Arrow stood it on its head: where others had accepted a particular binary collective-choice rule (Simple Majority Rule) and sought a reasonable way to generalize it to the multialternative case, Arrow accepted a particular way (BICH) to generalize any binary collective-choice rule to the multialternative case and sought a reasonable binary collective-choice rule that could be so generalized (dropping Feature 4 because Simple Majority Rule cannot be so generalized, as you saw in section 3.1).

An especially simple, widely instituted voting rule is *Plurality Rule:* Choose an alternative favored by the largest number of voters.

In the 1972 Chilean presidential election, there were three candidates, a leftist (l), a moderate (m), and a rightist (r). Although l got the most votes, none got a majority. So the election was thrown into the Chilean Congress, where the moderates had a plurality and the rightists were prepared to support m. But the moderates, less concerned to protect democracy than to be Good Democrats, thought the "popular choice" should be president and dogmatically assumed that Plurality Rule is the correct procedure for reckoning the popular choice. As a result, they supported l, who became president.

Was *l* really the popular choice? Was Plurality Rule the best procedure for picking the winner? Taken together, the moderate and rightist voters constituted a large majority, almost all of whom preferred *m* to *l*. Taken together, the moderate and leftist voters constituted an overwhelming majority, most of whom preferred *m* to *r*. So *m* was preferred to each of the other candidates by a majority. Were the candidates pitted against one another two at a time, *m* would have beaten both *l* and *r*. In short, *m* was best under the majority-preference relation. To put the point another way, no majority opposed *m*, but each of the other candidates was opposed by some majority.

Often Plurality Rule is modified as follows: When no feasible alternative is favored by a certain number or fraction of the voters, usually a majority, a *runoff* (second scrutiny) is held between the top two alternatives (the two favored by the greatest number of voters), or the top three, or some such thing. Depending on the number of runoff contestants, two or more successive runoffs might be held until the field can be narrowed no further. One can contrive any number of runoff rules—rules that require one or more runoff elections in specified circumstances.

When voting is *preferential*—when voters expressly rank the feasible alternatives instead of just specifying their favorites—one can reckon the result of a runoff, without actually holding a second round of voting, by deleting from everyone's ballot all alternatives but the runoff contestants and then reapplying Plurality Rule. When the alternatives eliminated are always those ranked first by the fewest voters, this procedure is sometimes called the *Alternative Vote Rule*, also the *Ware Rule*.

But when there are just three feasible alternatives, the only way to hold a runoff, or to delete alternatives ranked first by relatively few voters, is to pit the top two candidates against each other. And in the Chilean election, the top two candidates were *l* and *r*; *m*, the candidate majority-preferred to each of the others, would have been eliminated under such a procedure.

Another variation on the Plurality theme is the *Approval Voting Rule*, under which every voter lists one or more alternatives on his ballot but does not rank-order them, and the alternative with the greatest number of votes (the one listed on the greatest number of ballots) is chosen.

Normally, I suppose, a voter would list on his ballot all those alternatives ranked at or above some chosen level in his preference ordering—those he ranks first, or first plus second, or first plus second plus third, or the like.

Suppose the Approval Voting Rule is applied to four alternatives and six voters with the following preferences among those alternatives:

Mr. 1	Mr. 2	Mr. 3	Mr. 4	Mr. 5	Mr. 6
x	x	x	y	y	w
z	z	y	z	w	z
y	y	w	x	x	x
w	w	z	w	z	y

And suppose each voter votes for his first and second choices and nothing else. Then x and y receive three votes each, z receives four, and w receives two. So z wins. But that seems wrong. Like m in the previous example, x is preferred to each of the other feasible alternatives (including z) by a majority.

According to the *Borda Rule*, occasionally used in small organizations, the choice set comprises those feasible alternatives with a maximum Borda count relative to the feasible set, where the *Borda count* of x relative to α is the sum of x's rank numbers relative to α for all voters, and where x's *rank number* relative to α for Mr. i is a measure of how high Mr. i ranks x relative to the other alternatives in α: if x occurs at the bottom of his ranking of α, its rank number for him is 0; if x occurs one step up from the bottom, its rank number for him is 1; two steps up, 2; and so forth.

Suppose four voters rank five alternatives as follows:

Mr. 1	Mr. 2	Mr. 3	Mr. 4
x	x	x	y
y	y	y	z
z	z	z	w
w	w	w	v
v	v	v	x

Then x, y, z, w, and v have the following rank numbers and Borda counts relative to $\{x, y, z, w, v\}$:

	Rank Numbers				
	Mr. 1	Mr. 2	Mr. 3	Mr. 4	Borda Count
x	4	4	4	0	12
y	3	3	3	4	13
z	2	2	2	3	9
w	1	1	1	2	5
v	0	0	0	1	1

Since y has the largest Borda count relative to $\{x, y, z, w, v\}$, the Borda Rule prescribes the choice of y from this set. Yet x is preferred to every other alternative in $\{x, y, z, w, v\}$, including y, by a majority, indeed by one and the same majority, comprising all but one voter.

A feasible alternative majority-preferred to every other is called the *Condorcet alternative*. The trouble with the Plurality, runoff, Alternative Vote, Approval Voting, and Borda Rules is that sometimes there is a Condorcet alternative that is not chosen by these rules, contrary to:

CONDORCET CONDITION (COCO) *If* $x \in \alpha \in S$ *and* xMy *for every* $y \neq x$ *in* α *then* $C(\alpha) = \{x\}$.

We cannot simply adopt the rule that prescribes the choice of the Condorcet alternative. For there might *not exist* such an alternative. One reason is that there sometimes are tie votes. We can get around *that* problem by prescribing the choice of any feasible alternative to which none is majority-preferred. Either way, we still end up with no collective choice in Condorcet's Voting Paradox example and other cases.

A number of voting rules have been proposed that satisfy COCO yet apply in all cases. Here are two:

First, *Nanson's Rule*, a modification of the Borda Rule: Reckon the Borda count for all the feasible alternatives and eliminate those whose Borda counts are less than the mean Borda count, then do the same to the reduced feasible set, and continue in this way until further eliminations are impossible. Nanson (1883) proved that this rule satisfies COCO.

Suppose six voters rank x, y, z, v, and w as follows:

Mr. 1	Mr. 2	Mr. 3	Mr. 4	Mr. 5	Mr. 6
x	*x*	*y*	*y*	*w*	*z*
z	*z*	*w*	*w*	*v*	*x*
v	*w*	*v*	*v*	*z*	*v*
y	*y*	*z*	*z*	*x*	*y*
w	*v*	*x*	*x*	*y*	*w*

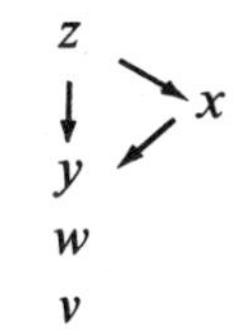

The accompanying diagram depicts the relation $M/\{x, y, z, w, v\}$ of majority preference among *x, y, z, w*, and *v*. In this example, only *v* and *y* have less than the mean Borda count, which is 12. For the reduced set, $\{x, z, w\}$, *x* alone has a Borda count less than the new mean, which is 6. And for $\{z, w\}$, the two alternatives share a Borda count of 3. Hence, Nanson's Rule allows the choice of *w* as well as *z*. But *z* is the only alternative of the five to which none is majority-preferred, and *z* is majority-preferred to *x* and *y*. That *w* should be choosable along with *z* seems anomalous.

Second, *Copeland's Rule:* Choose a feasible alternative with a maximum Copeland index relative to the feasible set, the *Copeland index* of *x* relative to α being the number of alternatives in α to which *x* is majority-preferred minus the number majority-preferred to *x*. This rule obviously satisfies COCO.

Suppose four voters rank *x, y, z, w*, and *v* as follows:

Mr. 1	Mr. 2	Mr. 3	Mr. 4
x	*x*	*y*	*v*
y	*y*	*w*	*z*
v	*z*	*z*	*w*
w	*w*	*v*	*x*
z	*v*	*x*	*y*

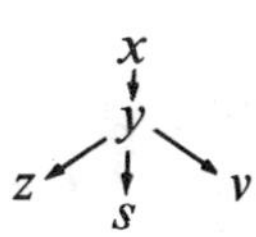

Since only *x* is majority-preferred to *y* but *y* is majority-preferred to three things in $\{x, y, z, w, v\}$, *y*'s Copeland index relative to this set is $3 - 1 = 2$. *x* has the Copeland index 1 ($=1 - 0$); and *z, w*, and *v* share a Copeland index of -1. So the Copeland Rule prescribes the choice of *y*. Yet *x* is majority-preferred to *y*, nothing is majority-preferred to *x*, and *x* alone enjoys the latter distinction.

Although Nanson's Rule and Copeland's satisfy COCO, what we have just seen is that both violate a close relative of COCO:

COCO 2 *If* $y \in \alpha \in S$, *if* x *is uniquely an M-undominated element of* α, *and if* xMy, *then* $y \notin C(\alpha)$.

A voting rule designed to satisfy not just COCO but COCO 2 is *Fishburn's Rule:* When α is the feasible set, choose an F^{α}-undominated element of α, where: $xF^{\alpha}y$ iff everything in α that is majority-preferred to x is majority-preferred to y and something in α is majority-preferred to y but not to x.

Let $\alpha = \{x, y, z, w\}$, and suppose six voters rank α thus:

Mr. 1	Mr. 2	Mr. 3	Mr. 4	Mr. 5	Mr. 6
x	x	y	y	z	z
w	y	z	z	w	w
y	w	x	x	x	x
z	z	w	w	y	y

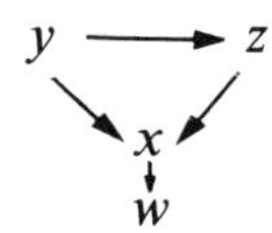

Among x, y, z, and w, which bear F^{α} to which? Consider each pair in turn: the one alternative (x) majority-preferred to y is not majority-preferred to x, and the one alternative (z) majority-preferred to x is not majority-preferred to y, so neither $xF^{\alpha}y$ nor $yF^{\alpha}x$. Similarly, neither $yF^{\alpha}z$ nor $zF^{\alpha}x$, neither $zF^{\alpha}x$ nor $xF^{\alpha}z$, neither $wF^{\alpha}z$ nor $zF^{\alpha}w$, and neither $xF^{\alpha}w$ nor $wF^{\alpha}x$. Finally, there is nothing majority-preferred to y but not to w and nothing majority-preferred to w but not to y, so neither $wF^{\alpha}y$ nor $yF^{\alpha}w$. Hence, nothing in α bears F^{α} to anything else in α. As a result, the Fishburn choice set for α is α itself. So Fishburn's Rule allows the choice of w from α. That seems unreasonable: w is foreign to the unique top M-cycle, one of whose elements is majority-preferred to w. This condition is violated:

COCO 3 *If* $x \in \alpha \in S$, *if some M-undominated element of* α *or an element of some top M-cycle in* α *bears M to* x, and *if* x *bears M to nothing in* α, *then* $C(\alpha) \neq \alpha$.

The intuitive idea behind COCO and its cousins is that an acceptable collective choice should, at least in some generalized sense, be a best choice, with majority preference the measure of better to worse. A particularly general form of this requirement is:

STRONG COCO *For* $\alpha \in S$, $C(\alpha)$ *is a subset of the union of minimum M-undominated subsets of* α.

This requires the collective choice always to belong to the GOCHA set under Simple Majority Rule. It does *not* require that $C(\alpha)$ *exhaust* that set. Obviously, Strong COCO implies COCO, COCO 2, and COCO 3.

One rule that certainly satisfies Strong COCO is the *GOCHA Rule* (sometimes called Schwartz's Rule): When α is the feasible set, choose anything in the union of minimum M-undominated subsets of α.

A problem with this rule (as I noted, in effect, in section 6.4) is that it violates:

PARETO EFFICIENCY *Suppose* $x, y \in \alpha \in S$, xP_iy *for some* $i \in N$, and yP_ix *for no* $i \in N$. *Then* $y \notin C(\alpha)$.

Say three voters rank four things this way:

Mr. 1	Mr. 2	Mr. 3
x	y	z
y	z	w
z	w	x
w	x	y

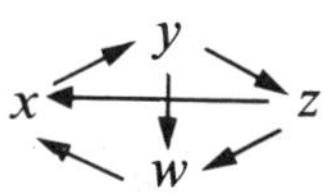

Then $\{x, y, z, w\}$ is the one and only minimum M-undominated subset of $\{x, y, z, w\}$. Therefore, the GOCHA Rule allows the choice of w from this set. Yet w is Pareto inefficient: all three voters prefer z to w.

8.2. THE SOLUTION

It is not reasonable or fair or desirable to make a choice by taking a vote in every circumstance. And very different voting rules could be reasonable or fair or desirable for different circumstances. What do they seek who seek a *reasonable* voting rule with Features 1–4?

One thing often sought, I think, particularly by those who expect a reasonable voting rule to satisfy COCO and its ilk, is a rule for *reckoning the majority will*—a collective-choice procedure that enforces *rule by majorities.* A bit more precisely: What is sought is a rule for determining, on the basis of voting information, *which feasible alternatives could be chosen if majorities and they alone wielded absolute power.*

Why not just present Messrs. 1, 2, . . . , n with the feasible alternatives, give majorities the power to dictate every choice, do everything necessary to ease communication and to ensure fidelity, and see which alternative is chosen? In short, why not play the majority cooperative game with the feasible alternatives as outcomes and see what happens?

Because such a procedure is very costly, especially if the chance of error is minimized. A voting rule of the sort I have described enables us to decide, at little cost, what would happen *if* the costly majority-game procedure were carried out, saving us the cost of carrying it out.

Let C represent a collective-choice process in which majorities wield absolute power and nonmajorities are impotent. Then, I contend, C must satisfy SOCO 2 with POW interpreted as the majoritarian power structure:

$$POW(g, x, y, \alpha) \text{ iff } g \subseteq N, x, y \in S, \text{ and } |g| > \tfrac{1}{2}|\{i \mid xP_iy \text{ or } yP_ix\}|.$$

This means that E_1, SUF, and E_2 are identical to the relations ME_1, SUM, and ME_2, and that D_2 and C are identical to the functions MAD and $MAMA$ (for *Multialternative Majority* Rule), all defined as follows:

DEFINITION $ME_1(g, x, y, \alpha)$ *iff* $g \subseteq N$, $x, y \in \alpha \in S$, xP_gy, *and* $|g| > |\{i \mid yP_ix\}|$.

DEFINITION $SUM(g, x, y, \alpha)$ *iff for no integer k do there exist* $z_0, z_1, \ldots, z_k, h_1, \ldots, h_k$ *such that*

(a) $z_0 = y \neq x = z_1$;
(b) $h_1 = g$;
(c) $ME_1(h_i, z_i, z_{i-1}, \alpha)$, $i = 1, 2, \ldots, k$;
(d) $h_{i-1} \subseteq h_i$ *or* $h_{i-1} \cap h_i = \emptyset$, $i = 2, 3, \ldots, k$; *and*
(e) *there do not exist* h, z *for which* $ME_1(h, z, z_k, \alpha)$ *and either* $h_k \subseteq h$ *or* $h_k \cap h = \emptyset$.

DEFINITION $ME_2(g, x, y, \alpha)$ *iff for some m there exist* $x_0, x_1, \ldots, x_m, g_1, \ldots, g_m$ *such that*

(a) $x_0 = y \neq x = x_m$;
(b) $g_k = g$;
(c) $ME_1(g_i, x_i, x_{i-1}, \alpha)$, $i = 1, 2, \ldots, m$; *and*
(d) $SUM(g, x, x_{m-1}, \alpha)$.

DEFINITION MAD *is the function on S whose value,* MAD^α, *at each*

$\alpha \in S$ is the relation $\{(x, y) \mid ME_2(g, x, y, \alpha)$ *for some g, but* $ME_2(g, y, x, \alpha)$ *for no* $g\}$.

DEFINITION *MAMA is the function on S for which*
$\text{MAMA}(\alpha) \equiv \{x \in \alpha \mid yMAD^{\alpha}x \text{ for no } y \in \alpha\}$.

Suppose $g \subseteq N$ and $x, y \in a \in S$. Then $ME_1(g, x, y, \alpha)$ holds if the g's prefer x to y and are more numerous than those individuals who prefer y to x. $SUM(g, x, y, \alpha)$ holds if, were g to block the choice of y from α in favor of x and stick together, then g could support x: apart from infinite regresses, g cannot do better than x and no disjoint majority opposes x. $ME_2(g, x, y, \alpha)$ holds if g possesses the first in a chain of majority preferences leading from x down to y and g can support x (assuming g sticks together). $xMAD^{\alpha}y$ holds if there is a chain of majority preferences leading from x down to y and beginning with the preference of some majority who can support x, but no such chain from y back to x. And $x \in MAMA(\alpha)$ if x could be effectively supported, or at least unopposed, were majorities to wield absolute power—if the collective choice of x from α is consistent with rule by majorities.

In view of Theorem 7.3.2, $MAMA(\alpha)$ is never empty. By Theorem 4.3.1, majority-unopposed (M-undominated) feasible alternatives always belong to the *MAMA* choice set. By Theorem 7.4.2, $MAMA(\alpha)$ is always a subset of the union of minimum M-undominated subsets of α; that is, *MAMA* satisfies Strong COCO and therewith COCO, COCO 1, and COCO 2. And by Theorem 7.4.6, *MAMA* satisfies Pareto Efficiency. So the *MAMA* Rule is acceptable from both Paretian and generalized Condorcetian points of view.

To illustrate the way the *MAMA* Rule (and therefore SOCO 2) works, suppose three voters rank $\alpha = \{x, y, z, w, v\}$ as follows:

Mr. 1	Mr. 2	Mr. 3
x	y	w
y	v	z
v	z	x
z	x	y
w	w	v

Then the ME_1 relationships within α are the following:

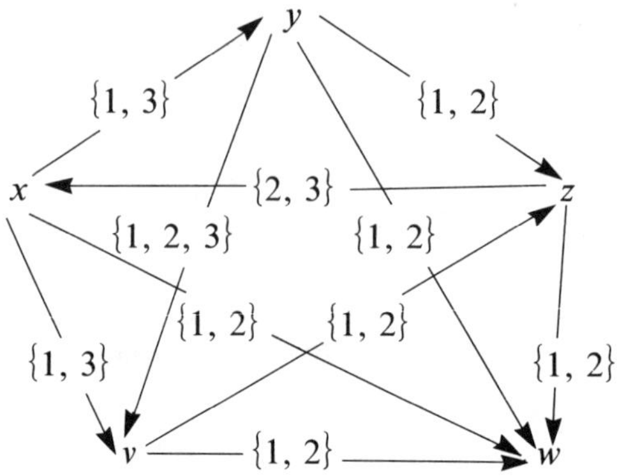

Note that there actually are four majorities who prefer y to v; in the diagram I listed only the most inclusive of them. Circles mark those ME_1 relationships for which *SUM* does not hold. The arrow from v to w, for example, has a circle because v is not the best $\{1, 2\}$ can do: Messrs. 1 and 2 would both be better off joining Mr. 3 to support y. The ME_2 relationships are given by all irreflexive chains of arrows that begin with uncircled arrows. Because the MAD^α relationships are all the one-way ME_2 relationships within α (or, equivalently, the one-way ME_2-chains within α) unrelativized to groups of individuals, MAD^α has the form

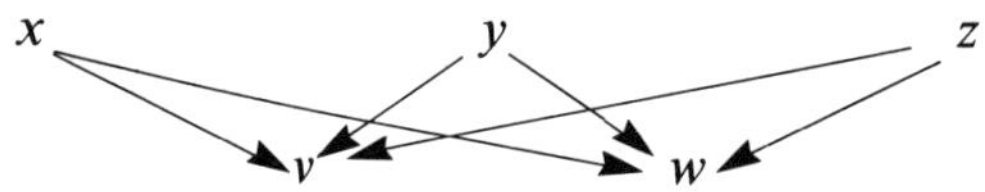

So $MAMA(\alpha) = \{x, y, z\}$. By contrast, with $M = P$, $C(\alpha) = \emptyset$ according to BICH, $C(\alpha) = \alpha$ according to GETCHA, and $C(\alpha) = \{x, y, z, v\}$ according to GOCHA.

Because everyone prefers y to v, Pareto Efficiency requires that v not belong to $C(\alpha)$. Each of the other four alternatives is Pareto efficient. In the example, then, $MAMA(\alpha)$ is the intersection of the GOCHA set and the Pareto set—the set of Pareto-efficient alternatives. As you know, $MAMA(\alpha)$ is always a subset of this intersection. Sometimes it is a *proper* subset; witness the following example:

Three voters rank $\alpha = \{x, y, z, w\}$ thus:

Mr. 1	Mr. 2	Mr. 3
x	*z*	*y*
w	*x*	*z*
y	*w*	*w*
z	*y*	*x*

Here are the ME_1 relationships within α:

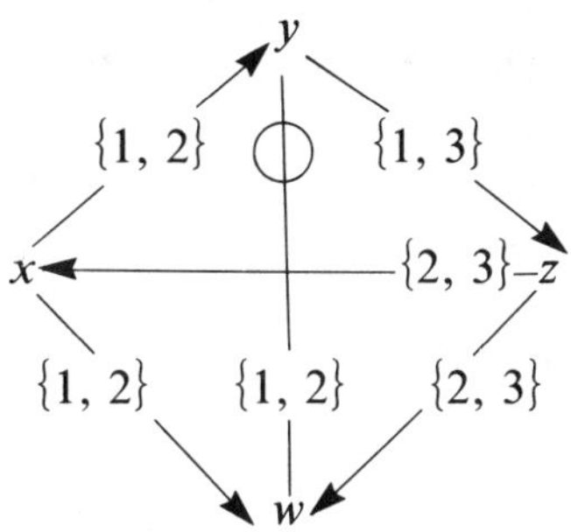

A circle marks the one ME_1 relationship for which *SUM* does not hold: because Messrs. 1 and 2 prefer *x* to *w*, *w* is not the best {1, 2} can do. So MAD^{α} has the following form:

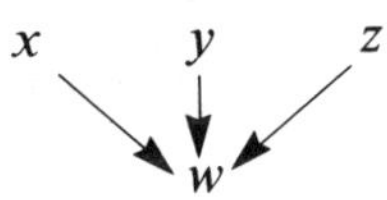

Because α is an *M*-cycle and every member is Pareto efficient, α itself is both the GOCHA set from α—assuming $M = P$—and the Pareto set. But *MAMA*(α) = {*x*, *y*, *z*}.

SOURCES AND RELATED CONTRIBUTIONS

GOCHA was first formulated in the final paragraph of Schwartz (1970), developed in detail in Schwartz (1972), and named in Schwartz (1977). The axiomatic characterization comes from Schwartz (1972). Deb (1977) provides the equivalent transitive-closure formulation and shows that GOCHA can conflict with Pareto Efficiency. I thought of GETCHA in 1968 but lost interest in it when, shortly thereafter, I discovered GOCHA; my interest was revived by the needs of chapter 9.

GETCHA is equivalent to a choice-set specification developed by Ferejohn and Grether (1977).

In its present form, SOCO 2 has not previously appeared in print. But a less refined version (equivalent to the present version when applied to majoritarian power structures) is mentioned in Schwartz (1977) and examined closely in Richelson (1981). I discuss SOCO 2 again at some length in chapter 11, where I show, among other things, that it yields the intuitively appropriate solution to Sen's famous Liberal Paradox. In part, it was the needs of chapter 11 that gave rise to this condition: I sought a solution concept what would work well for both majoritarian and decentralized (including libertarian) power structures.

Black (1958: Part II) traces the early history of the voting-rule problem. Borda (1781), Nanson (1882), Copeland (1951), and Fishburn (1977) propose the rules that here bear their names. Brams and Fishburn (1978) advocate the Approval Voting Rule. Other interesting proposals are made by Dodgson (Lewis Carrol) (1873, 1874, 1876), Rescher (1972), and Miller (1977a, 1980), who proposes a rule similar to Fishburn's. Gärdenfors (1973), Smith (1973), and Young (1974, 1975) treat the rationale behind the Borda Rule and other "positionalist" voting rules. Fishburn (1971a; 1973: chs. 12, 13; 1974a; 1977a) and Richelson (1975, 1978a, 1978b, 1978d, 1981) offer impressive systematic surveys and analyses of a large number of voting rules.

PART IV

COLLECTIVE-CHOICE PROCESSES DISSECTED

CHAPTER NINE

MULTISTAGE CHOICE PROCESSES

9.1. THE CONCEPT OF A MULTISTAGE CHOICE

Often a choice is made not in one fell swoop but in two or more stages: it is the final result of a series of choices, one from each of several jointly exhaustive subsets of the feasible set.

Example 1. Faced with a motion, an amendment, and a substitute motion, a committee votes on the amendment, then the substitute, then the surviving motion—original, amended, or substitute. In so doing it pits the original motion against the amendment, the winner against the substitute, and the winner in the latter contest against the status quo. The result is a three-stage choice from the set {status quo, motion, amendment, substitute}. The three stages can be represented in the tree diagram of figure 9.1.

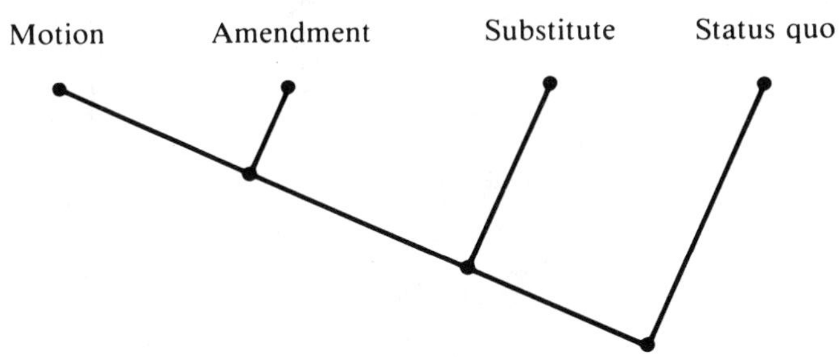

Figure 9.1

Example 2. A single diner or committee of banquet planners chooses a luncheon dish by choosing between two fish dishes and between two meat dishes, then comparing the chosen fish dish with the chosen meat dish; see figure 9.2.

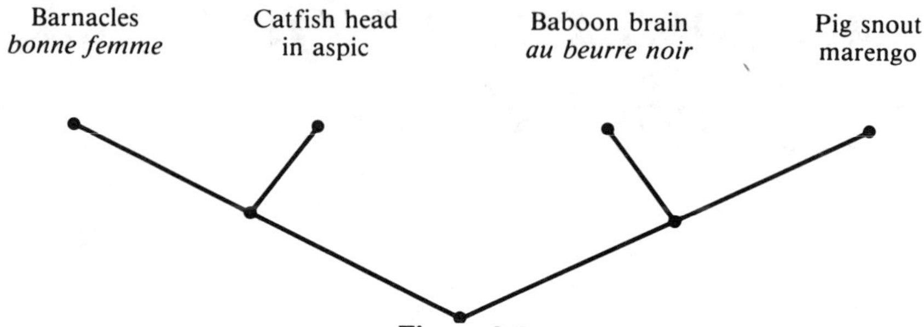

Figure 9.2

Example 3. Current government policy on some issue is the result of a four-stage evolution. At time t_1 a choice is made among the policy alternatives feasible then. This choice is treated as the status quo at t_2, when it is compared with one or more new options. The choice at t_2 becomes the new status quo, and it belongs to the feasible set at t_3. At t_4 the choice made at t_3 is pitted against some newer alternatives. At each stage, the choice belongs to the choice set from the feasible set at that stage; it is automatically the status quo if the status quo belongs to the choice set. See figure 9.3.

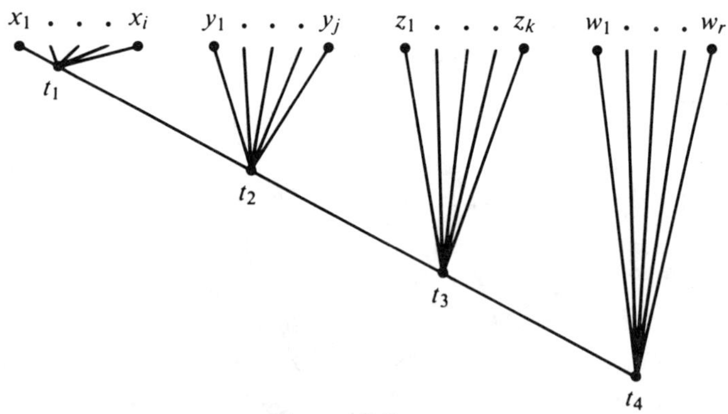

Figure 9.3

In this chapter I try to show precisely how multistage processes effect (or fail to effect) final outcomes.

Examples 1 and 2 are *binary* multistage choices: there are just two feasible alternatives at each stage. Examples 1 and 3 are *serial* choices: each successive stage pits the previous choice against one or more new alternatives until all the alternatives are exhausted.

Every multistage choice from a set α works this way: α is divided into two or more nonempty (possibly overlapping) subsets, which together exhaust α; an alternative is chosen from each subset, and the final choice is made from among the alternatives thus chosen. The choice from each subset is made the same way—by dividing that subset into nonempty, jointly exhaustive subsets, choosing from each, and choosing among the chosen alternatives—unless the given subset is a unit set.

Apart from limiting cases, there are several feasible sets involved in a multistage choice. There is the set whence the whole multistage choice is made. And there is another feasible set at each of the several stages in the process. The latter must be a subset of the former.

The result of a multistage choice depends, of course, on the choice process used at each stage. Hereafter, suppose the choice at each stage in any multistage choice is made according to C. So if α is the feasible set at some stage in some multistage choice, the choice at that stage will be one of the alternatives in $C(\alpha)$. (Although one can imagine the use of different choice functions at different stages, to allow this would be to create a picture vastly more complex and less tractable than the sufficiently complex one with which I am dealing.)

A choice set might have more than one member. Call a multi-member choice set a *tie*. You *break* a tie when you choose one of the several alternatives comprised therein. A procedure for doing this is a *tie-breaking procedure*. The final result of a multistage choice with ties along the way depends, in part, on the tie-breaking procedure used.

This result also depends, in part, on the way the feasible set is divided into subsets, the way these subsets are divided in turn into subsets, and so forth. A procedure for making such divisions is a *division procedure*. Its application to a set α can be depicted by a *tree diagram*. Each *node* represents some subset β of α, and its *successor* nodes—those, if any, into which it immediately branches—represent the subsets into which β is divided. The bottom node represents α itself. The top nodes (those devoid of successors) represent all the unit subsets of α.

Some might question my policy of representing every multistage choice in such a way that the process of successive division into subsets does not terminate until *unit sets* are reached. I have adopted this policy for technical convenience only. It imposes no real restriction on the choices represented. Suppose it would be natural to describe a choice from a large set α as the result of dividing α into two subsets, β and γ, choosing from each in one fell swoop, then choosing between the two chosen alternatives. Although this description says nothing about dividing β and γ into their respective unit subsets, I

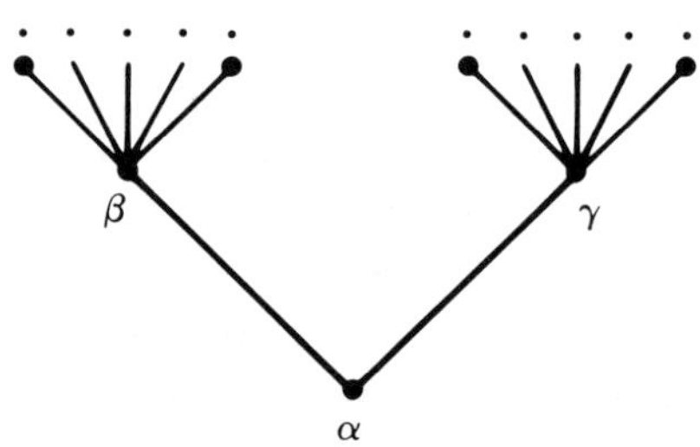

Figure 9.4

should still represent the choice thus described by the tree diagram of figure 9.4, in which the top nodes represent the unit subsets of β and γ, rather than that of figure 9.5.

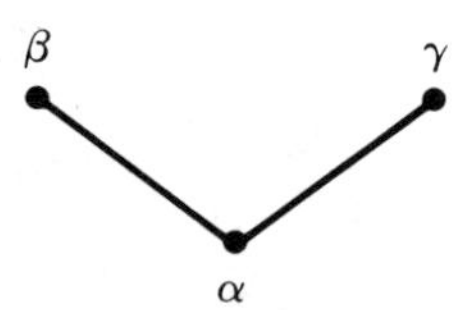

Figure 9.5

Mathematically speaking, just what is a *division procedure*? An obvious suggestion is that it is a relation—the relation of each set $\alpha \in S$ to each of those subsets into which the procedure divides α.

Another suggestion is that a division procedure is a function—that which transforms each set $\alpha \in S$ into the *sequence* of subsets into which the procedure divides α, the order of subsets in the sequence reflecting the left-to-right order of nodes in the corresponding tree diagram. Because the left-to-right order of tree nodes represents the temporal order (if any) of choices, the function approach captures more potentially important dendriform structure than the relation approach.

The two approaches share a shortcoming. There are imaginable division procedures that cannot be identified with relations or functions of the sort contemplated. Example: d is a division procedure that divides $\{x, y, z\}$ into $\{x, y\}$ and $\{z\}$, then $\{x, y\}$ into $\{x\}$ and $\{x, y\}$ itself, and finally $\{x, y\}$ into $\{y\}$ and $\{x\}$, as in figure 9.6. On the relation approach, because the first $\{x, y\}$-node branches into another $\{x, y\}$-node but the second does not, we should have both $\{x, y\}\rho\{x, y\}$ and not

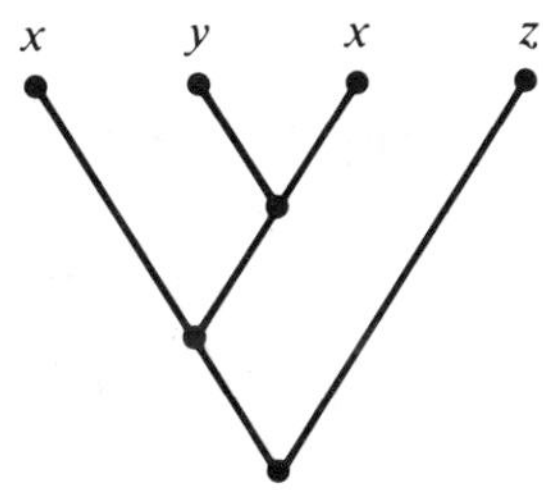

Figure 9.6

$\{x, y\}\rho\{x,y\}$—a contradiction. And on the function approach, because the first $\{x, y\}$-node branches into $\{x\}$- and $\{x, y\}$-nodes whereas the second branches into $\{y\}$- and $\{x\}$-nodes, we should have both $\phi(\{x, y\}) = (\{x\}, \{x, y\})$ and $\phi(\{x, y\}) = (\{y\}, \{x\})$, which is impossible if $x \neq y$.

This objection may not be devastating. Division procedures not representable by relations or functions in the suggested manner may be rare in practice. Still, it would be best to characterize multistage choice processes and therewith division procedures in a way that encompasses all conceivable ones.

My approach: A division procedure applies to a feasible set α by transforming α into a *sequence* s of alternatives, then dividing s into two or more subsequences, dividing these in turn into subsequences, and so on, until unit sequences are reached. Called the *agenda sequence* for the situation in which α is the feasible set, s matches the left-to-right order of top nodes in the corresponding tree diagram. The operation that divides sequences into subsequences is like the functions discussed above except that it applies to sequences rather than sets of alternatives and yields sequences of sequences rather than sequences of sets of alternatives. In the example that caused trouble for the other approaches, $\{x, y, z\}$ is transformed into the sequence (x, y, x, z) of alternatives, this sequence is transformed into the sequence $((x, y, x), (z))$ of sequences of alternatives, the nonunit component sequence (x, y, x) of alternatives is transformed into the sequence $((x), (y, x))$ of sequences, and (y, x) is transformed finally into $((y), (x))$.

So we can conceive of a division procedure as consisting of two things: First, a transformation of each $\alpha \in S$ into an α-sequence, where:

DEFINITION *s is an* α-sequence *iff* α *is a set and s a finite sequence whose terms are all and only the members of* α.

Second, a function that assigns, to every finite sequence s of alternatives, a sequence of mutually exclusive, jointly exhaustive subsequences of s; call it a *division function*.

Let *SEQ* be the set of potential agenda sequences.

DEFINITION *SEQ = the set of finite sequences whose terms all belong to A.*

If a sequence $s \in SEQ$ has two or more terms, a division function F divides s into two or more subsequences, say $s_1, \ldots, s_k$, so that $F(s) = (s_1, \ldots, s_k)$. Not just any sequence of subsequences of s can constitute a division of s. What makes $(s_1, \ldots, s_k)$ a division of s is that its *concatenation*, $CAT(s_1, \ldots, s_k)$, is identical to s. In general, $CAT(s_1, \ldots, s_k)$ is the sequence of alternatives consisting of the terms of s_1 followed by the terms of s_2 and so on. Example: $CAT((x, y), (z), (w, v, x, t)) = (x, y, z, w, v, x, t)$.

DEFINITION *CAT is the function defined on the set of finite sequences of members of SEQ by the following recursion:*

$$CAT((s)) = s$$

$$CAT((x_1, \ldots, x_i), (y_1, \ldots, y_j)) = (x_1, \ldots, x_i, y_1, \ldots, y_j)$$

$$CAT((s_1, \ldots, s_k, s_{k+1})) = CAT((CAT((s_1, \ldots, s_k)), s_{k+1})).$$

DEFINITION *A* division function *is a function F on SEQ into the family of finite sequences of members of SEQ such that, for every* $s \in SEQ$,
(a) *if s has two or more terms, so does* $F(s)$, and
(b) $s = CAT(F(S))$.

Let us represent any multistage choice process by a function Π of sequences of alternatives. If $s \in SEQ$, $\Pi(s)$ is the alternative chosen by the process in question when s is the agenda sequence. Although Π is a function of potential agenda sequences rather than potential feasible sets, there is a unique potential feasible set corresponding to every potential agenda sequence. It is the set comprising all and only the terms of the given sequences.

A function like Π is a *multistage choice operation*—a *MUSTACH operation*, for short. Different MUSTACH operations embody different division functions or tie-breaking procedures. A MUSTACH operation *relative to F* is one for which F is the underlying division function.

DEFINITION Π *is a* MUSTACH operation relative to F *iff* F *is a division function and* Π *is a function on SEQ such that*
(a) $\Pi((x)) = x$ *for every* $x \in A$ *and*
(b) *for every k and all* $s, s_1, \ldots, s_k \in SEQ$, *if* $F(s) = (s_1, \ldots, s_k)$ *then* $\Pi(s) \in C(\{\Pi(s_1), \ldots, \Pi(s_k)\})$.

DEFINITION Π *is a* MUSTACH operation *iff* Π *is a MUSTACH operation relative to something.*

To make a multi-stage choice from a set α, then, we need an α-sequence, $(x_1, \ldots, x_m)$, a division function, F, and a MUSTACH operation, Π, relative to F. The alternatives $x_1, \ldots, x_m$ are precisely the members of α, arranged in one particular order. F divides $(x_1, \ldots, x_m)$ into m or fewer subsequences, divides each of these in turn into subsequences, and so on, until unit sequences are reached; say $F(x_1, \ldots, x_m) = (s_1, \ldots, s_k)$. Note that each x_i is an *alternative* whereas each s_j is a *sequence* of alternatives. $\Pi(x_1, \ldots, x_m)$, the final result of the multi-stage choice, is one of the alternatives x_i, and it is reckoned by first reckoning $\Pi(s_1), \ldots, \Pi(s_k)$ and then reckoning $C(\{\Pi(s_1), \ldots, \Pi(s_m)\})$. If there is a tie—if $C(\{\Pi(s_1), \ldots, \Pi(s_m)\})$ contains more than one alternative—then Π somehow breaks the tie: $\Pi(x_1, \ldots, x_m)$ is just one of the members of $C(\{\Pi(s_1), \ldots, \Pi(s_k)\})$.

To illustrate, suppose a motion, an amendment, and a substitute motion have been moved and seconded in a three-member committee that uses Majority Rule. And suppose the members have (and are willing to express) the following preferences among the original motion (m), its amended version (a), the substitute (s), and the status quo (q):

Mr. 1	Mr. 2	Mr. 3
m	a	s
a	q	q
q	s	m
s	m	a

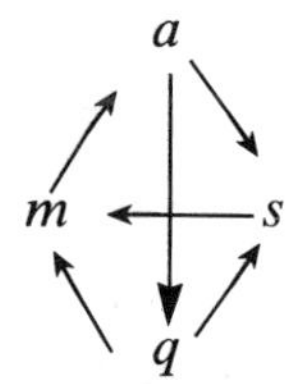

Under standard parliamentary procedure, m would be pitted against a, the winner against s, and the winner in the latter contest against q. This procedure may be described in terms of the agenda sequence (m, a, s, q) and a division function F and MUSTACH operation Π for which $F(m, a, s, q)$ and $\Pi(m, a, s, q)$ are reckoned as follows:

$$F(m, a, s, q) = ((m, a, s), (q))$$
$$F(m, a, s) = ((m, a), (s))$$
$$F(m, a) = ((m), (a))$$

$$\Pi(m, a) \in C(\{m, a\}) = \{m\}$$
$$\Pi(m, a, s) \in C(\{\Pi(m, a), s\}) = C(\{m, s\}) = \{s\}$$
$$\Pi(m, a, s, q) \in C(\{\Pi(m, a, s), q\}) = C(\{s, q\}) = \{q\}.$$

So q is the final choice. Figure 9.7 is the tree diagram. Alternatively, we could have described the same situation using the agenda sequence (q, m, a, s) and a division function G for which $G(q, m, a, s) = ((q), (m, a, s))$. The tree diagram then would be as in figure 9.8. Either way, had s instead been moved (and perhaps reworded) as an amendment and a as a substitute motion, their order in the agenda sequence would have been reversed, and the final choice would have been a rather than q.

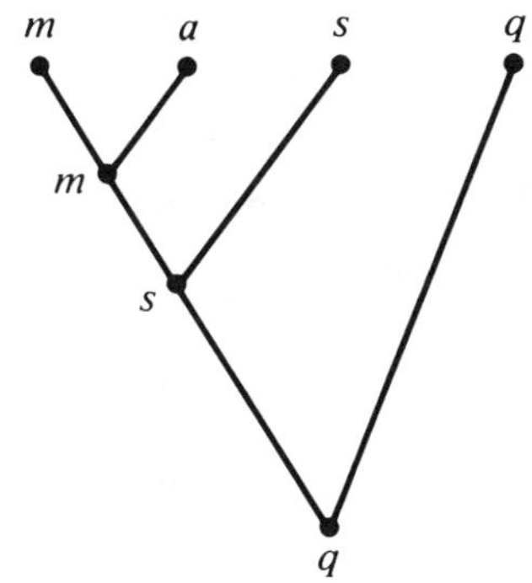

Figure 9.7

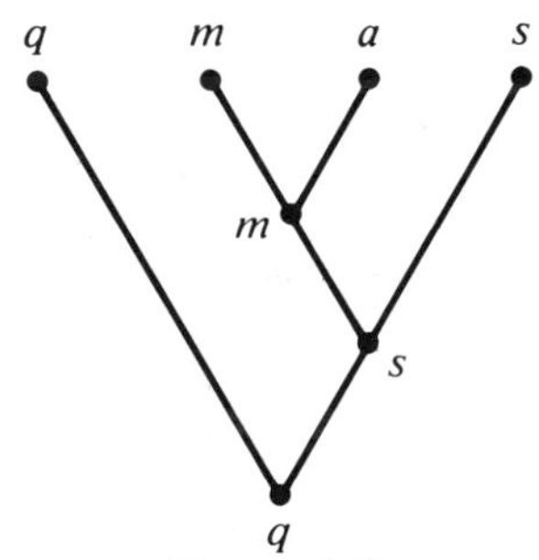

Figure 9.8

9.2. SOME SALIENT SORTS OF MULTISTAGE CHOICE PROCESS

Construed simply as a member of *SEQ*, a potential agenda sequence can be *redundant* in this sense:

DEFINITION *A* redundant sequence *is a sequence in which some term occurs more than once.*

Thus, (1, 2, 1) is redundant but (1, 2, 3) is not. Often there are institutional restrictions against redundant agenda sequences. Often a motion cannot come up for a vote more than once in the same legislative or committee session. For such situations, the potential agenda sequences are not all the members of *SEQ*, just the non-redundant ones.

I gave examples of binary and serial multistage choices, including one that was both binary and serial. Whether a multistage choice or choice process is binary and whether it is serial depend on the underlying division function.

In a binary multistage choice, the choice at each stage involves just two alternatives. So a binary division function divides each sequence consisting of two or more alternatives into exactly two subsequences.

DEFINITION *A* binary division function *is a division function F such that, for every* $s \in SEQ$, *if s has two or more terms then F(s) has exactly two terms.*

DEFINITION Π *is a* binary MUSTACH operation relative to *F iff* Π *is a MUSTACH operation relative to F, and F is a binary division function.*

In a serial choice, each successive stage pits the choice made at the previous stage against one or more new alternatives. So the agenda sequence is divided into an initial subsequence, s, plus one or more unit sequences, $(x_1), \ldots, (x_k)$. The penultimate choice, say x, is made from s, and the final choice is made from $(x, x_1, \ldots, x_k)$. s is divided the same way, and the division process continues until unit sequences are reached.

DEFINITION *F is a* serial division function *iff F is a division function, and for every k and all* $s, s_1, \ldots, s_k \in SEQ$, $F(s) = (s_1, s_2, \ldots, s_k)$ *only if* $s_2, \ldots, s_k$ *are unit sequences.*

DEFINITION Π *is a* serial MUSTACH operation relative to *F iff* Π *is a MUSTACH operation relative to F, and F is a serial division function.*

Another interesting type of multistage choice, illustrated by Example 2, is *choice by division and conquest:* The feasible set is partitioned, a choice is made in one fell swoop from each set belonging to the partition, then a choice is made among the alternatives thus chosen.

An important difference among MUSTACH operations consists in the tie-breaking procedures they embody. A natural procedure for breaking ties, especially in serial choices, is the Conservative Rule: Choose the previously chosen alternative if it belongs to the tie. In other words, stick with the status quo when possible. This is only a partial tie-breaking procedure. It is not applicable to all ties. In a binary, serial choice, every tie involves just two alternatives, one of which must be the previous choice (counting the first alternative in the agenda sequence as the first choice). But in nonbinary choices, even serial ones, a tie need not contain the previous choice.

DEFINITION Π *is a* conservative MUSTACH operation relative to F *iff* Π *is a MUSTACH operation relative to F, and for every k and all s,* $s_1, \ldots, s_k \in SEQ$, *if* $F(s) = (s_1, \ldots, s_k)$ *and* $\Pi(s_1) \in C(\{\Pi(s_1), \ldots, \Pi(s_k)\})$ then $\Pi(s) = \Pi(s_1)$.

Some tie-breaking procedures are *exogenous* in the sense that the way they break ties depends only on the tied alternatives, not on the agenda sequence or division function. The Conservative Rule is not exogenous: how it breaks a tie depends on which alternative was previously chosen. A familiar exogenous tie-breaking procedure is the Chairman's Rule: Choose the tied alternative favored by the chairman. Any strictly *random* procedure for breaking ties also is exogenous.

Under some exogenous tie-breaking procedures (under *every* such procedure designed for *binary* choices), there exists an asymmetric binary relation ρ—a *tie breaker*, let us call it—such that ties are broken in favor of the ρ-first tied alternative—the tied alternative that bears ρ to every other tied alternative. ρ might represent a series of coin tosses. It also might represent a chairman's preferences.

DEFINITION ρ *is a* tie breaker for Π relative to F *iff* Π *is a MUSTACH operation relative to F,* ρ *is an asymmetric binary relation on A, and for every k and all s,* $s_1, \ldots, s_k \in SEQ$, *if* $F(s) = (s_1, \ldots, s_k)$ *then* $\Pi(s)$ *is the* ρ*-first element of* $C(\{\Pi(s_1), \ldots, \Pi(s_k)\})$.

Among tied alternatives, there might be none that bears a given binary relation, even an asymmetric one, to every other. But there must be such an alternative if the relation is a *linear ordering* of *A*.

With *C* fixed, a binary division function by itself uniquely determines a conservative MUSTACH operation, and a division function plus linear tie breaker together uniquely determine a MUSTACH operation. That is, if *F* is any division function and ρ any linear ordering of A, there is a unique conservative MUSTACH operation relative to F if F is binary, and there is a unique MUSTACH operation relative to F for which ρ is a tie breaker.

Under *normal* multistage choice processes, the tie-breaking procedure depends only on the tied alternatives and their order in the agenda sequence—not, for example, on the history of prior or future stages in the process.

DEFINITION Π *is a* normal MUSTACH operation relative to F *iff* Π *is a MUSTACH operation relative to F, and for every k and all s,* $s_1, \ldots, s_k \in SEQ$, *if* $F(s) = (s_1, \ldots, s_k)$ then $\Pi(s) = \Pi(\Pi(s_1), \ldots, \Pi(s_k))$.

It is obvious that binary, conservative MUSTACH operations and MUSTACH operations with tie breakers are normal.

9.3. THE INDETERMINACY OF MULTISTAGE CHOICE PROCESSES

Just which alternatives can be chosen from a set α by a multistage process?

The final result of a multistage choice from α depends, of course, on α and C. It also depends on the agenda sequence, division function, and tie-breaking procedure used. That the result depends on α and C is expected. The choice is *from* α, and C represents the will, the preferences, the values of those doing the choosing. That the result depends *to some extent* on the agenda sequence, division function, and tie-breaking procedure is perhaps less expected but no shock. What you will find surprising, I think, is the *extent* to which the result depends on these last three, seemingly arbitrary factors. With C and α fixed, the set of alternatives choosable from α by a multistage process is surprisingly large.

Multistage Choice and GOCHA

An alternative in α that bears P to every other alternative in α can be the result of a multistage choice from α. So, more generally, can any P-undominated element of α—any member of the BICH set from α. So, indeed, can any member of any minimum P-undominated subset of α—any member of the GOCHA set from α. That is:

Suppose $\alpha \in S$ and x belongs to some minimum P-undominated subset of α. Then there is an α-sequence s and a MUSTACH operation Π for which $\Pi(s) = x$.

This result can be strengthened a great deal. It does not just hold for some α-sequence or other. It holds for some *nonredundant* one. It does not just hold for some MUSTACH operation or other. It holds for some *binary, serial, conservative* one. It holds as well for some binary, serial MUSTACH operation with a *linear tie breaker*. Anything choosable from α according to GOCHA can be chosen from α by means of a multistage choice process of the simplest sort.

To prove this, I shall need two lemmata.

LEMMA 9.3.1 *If Π is a normal, binary, serial MUSTACH operation relative to F and if $x_1, \ldots, x_i, x_{i+1}, \ldots, x_k \in A$ then $\Pi(x_1, \ldots, x_i, x_{i+1}, \ldots, x_k) = \Pi(\Pi(x_1, \ldots, x_i), x_{i+1}, \ldots, x_k)$.*

Proof. By induction on k. Because F is binary and serial,

$$F(x_1, \ldots, x_i, x_{i+1}, \ldots, x_k) = ((x_1, \ldots, x_i, x_{i+1}, \ldots, x_{k-1}), (x_k))$$

and

$$F(\Pi(x_1, \ldots, x_i), x_{i+1}, \ldots, x_k) = ((\Pi(x_1, \ldots, x_i), x_{i+1}, \ldots, x_{k-1}), (x_k)).$$

Therefore, because Π is normal relative to F,

$$\begin{aligned}\Pi(x_1, \ldots, x_i, x_{i+1}, \ldots, x_k) &= \Pi(\Pi(x_1, \ldots, x_i, x_{i+1}, \ldots, x_{k-1}), \Pi(x_k)) \\ &= \Pi(\Pi(\Pi(x_1, \ldots, x_i), x_{i+1}, \ldots, x_{k-1}), \Pi(x_k)) \\ &\qquad \text{by inductive hypothesis} \\ &= \Pi(\Pi(x_1, \ldots, x_i), x_{i+1}, \ldots, x_{k-1}, x_k) \\ &\qquad \text{because } \Pi \text{ is normal.} \blacksquare\end{aligned}$$

LEMMA 9.3.2. *Suppose:*

(1) $x, \ldots, x_k, w_1, \ldots, w_r \in A$, $k \geq 1$, $r \geq 0$;

(2) $(y_1, \ldots, y_m)$ *is the result of deleting from* $(x_1, \ldots, x_k)$ *all but the last occurrence of each* x_i *that occurs more than once;*

(3) Π *is a normal, serial, binary MUSTACH operation relative to* F;

(4) $\Pi(x_k, w_i) = x_k$, $i = 1, 2, \ldots, r$; *and*

(5) $\Pi(x_i, x_{i+1}) = x_{i+1}$, $i = 1, 2, \ldots, k - 1$.

Then $\Pi(y_1, \ldots, y_m, w_1, \ldots, w_r) = x_k$.

Proof. By induction on $k + r$. By construction,

$y_m = x_k$.

Suppose $r \geq 1$. Then by Lemma 9.3.1,

$$\begin{aligned}\Pi(y_1, \ldots, y_m, w_1, \ldots, w_r) &= \Pi(\Pi(y_1, \ldots, y_m, w_1, \ldots, w_{r-1}), w_r)\\ &= \Pi(x_k, w_r) \qquad \text{by inductive hypothesis}\\ &= x_k \qquad \text{by (4).}\end{aligned}$$

Now suppose $r = 0$; to prove that $\Pi(y_1, \ldots, y_m) = x_k$. Let $(z_1, \ldots, x_p)$ be the result of deleting from $(x_1, \ldots, x_{k-1})$ all but the last occurrence of each x_i that occurs more than once. Then

$x_p = x_{k-1}$.

Three cases:

Case 1. $x_k \in \{z_1, \ldots, z_p\}$. Then

$(y_1, \ldots, y_m) = (z_1, \ldots, z_p, x_k)$.

So

$$\begin{aligned}\Pi(y_1, \ldots, y_m) &= \Pi(z_1, \ldots, z_p, x_k)\\ &= \Pi(\Pi(z_1, \ldots, z_p), x_k) \qquad \text{by Lemma 9.3.1}\\ &= \Pi(x_{k-1}, x_k) \qquad \text{by inductive hypothesis}\\ &= x_k \qquad \text{by (5).}\end{aligned}$$

Case 2. $x_k = z_1$. Then

$(y_1, \ldots, y_m) = (z_2, \ldots, z_p, x_k)$.

So

$$\begin{aligned}\Pi(y_1, \ldots, y_m) &= \Pi(z_2, \ldots, z_p, x_k)\\ &= \Pi(\Pi(z_2, \ldots, z_p), x_k) \qquad \text{by Lemma 9.3.1}\\ &= \Pi(x_{k-1}, x_k) \qquad \text{by inductive hypothesis}\\ &= x_k \qquad \text{by (5).}\end{aligned}$$

Case 3. for some $i = 1, 2, \ldots, p - 1$, $x_k = z_{i+1}$. Then

$(z_1, \ldots, z_p) = (z_1, \ldots, z_{i-1}, x_k, z_{i+1}, \ldots, z_p)$

and

$$(y_1, \ldots, y_m) = (z_1, \ldots, z_{i-1}, z_{i+1}, \ldots, x_p, x_k).$$

Two subcases:

Subcase 3.1. There is a $j = i + 1, \ldots, p$ for which

$$z_j = \Pi(z_1, \ldots, z_{i-1}, z_{i+1}, \ldots, z_j).$$

But

$$\begin{aligned}
x_k &= \Pi(x_{k-1}, x_k) && \text{by (5)} \\
&= \Pi(\Pi(z_j, \ldots, z_p), x_k) && \text{by inductive hypothesis} \\
&= \Pi(z_j, \ldots, x_p, x_k) && \text{by Lemma 9.3.1} \\
&= \Pi(\Pi(z_1, \ldots, z_{i-1}, z_{i+1}, \ldots, z_j), z_{j+1}, \ldots, z_p, x_k) \\
& && \text{by hypothesis of the Subcase} \\
&= \Pi(z_1, \ldots, z_{i-1}, z_{+1}, \ldots, z_j, \ldots, z_p, x_k) && \text{by Lemma 9.3.1} \\
&= \Pi(y_1, \ldots, y_m).
\end{aligned}$$

Subcase 3.2. There is *no* $j = i + 1, \ldots, p$ for which $\Pi(z_1, \ldots, z_{i-1}, z_{i+1}, \ldots, z_j) = z_j$. But by Lemma 9.3.1, for every $j = i + 1, \ldots, p$, $\Pi(z_1, \ldots, z_{i-1}, z_{i+1}, \ldots, z_j) = \Pi(\Pi(z_1, \ldots, z_{i-1}, z_{i+1}, \ldots, z_{j-1}), z_j)$, so that $\Pi(z_1, \ldots, z_{i-1}, z_{i+1}, \ldots, z_j)$ is either $\Pi(z_1, \ldots, z_{i-1}, z_{i+1}, \ldots, z_{j-1})$ or z_j. Thus, by hypothesis of the Subcase,

$$\Pi(z_1, \ldots, z_{i-1}, z_{i+1}, \ldots, z_j) = \Pi(z_1, \ldots, z_{i-1}, z_{i+1}, \ldots, z_{j-1}), \quad j = i + 1, \ldots, p.$$

By $j - i$ successive applications of this consequence, we obtain

$$\Pi(z_1, \ldots, z_{i-1}, z_{i+1}, \ldots, z_j) = \Pi(z_1, \ldots, z_{i-1}), \ j = i + 1, \ldots, p.$$

In particular, then,

$$(*) \quad \Pi(z_1, \ldots, z_{i-1}, z_{i+1}, \ldots, z_p) = \Pi(z_1, \ldots, z_{i-1}).$$

But since, by construction, $(z_1, \ldots, z_p)$ is nonredundant,

$$\Pi(z_1, \ldots, z_{i-1}) \neq z_p = x_{k-1}.$$

By inductive hypothesis, however,

$$\Pi(z_1, \ldots, z_{i-1}, x_k, z_{i+1}, \ldots, z_p) = \Pi(z_1, \ldots, z_p) = x_{k-1}.$$

Consequently,

$$\Pi(z_1, \ldots, z_{i-1}, x_k, z_{i+1}, \ldots, z_p) \neq \Pi(z_1, \ldots, z_{i-1}, z_{i+1}, \ldots, z_p)$$

whence, by Lemma 9.3.1,

$$\Pi(\Pi(z_1, \ldots, z_{i-1}, x_k), z_{i+1}, \ldots, z_p) \neq \Pi(\Pi(z_1, \ldots, z_{i-1}), z_{i+1}, \ldots, z_p)$$

so that

$$\Pi(z_1, \ldots, z_{i-1}, x_k) \neq \Pi(z_1, \ldots, z_{i-1})$$

and thus, by Lemma 9.3.1 again,

$$\Pi(\Pi(z_1, \ldots, z_{i-1}), x_k) \neq \Pi(z_1, \ldots, z_{i-1}).$$

But $\Pi(\Pi(z_1, \ldots, z_{i-1}), x_k)$ is either $\Pi(z_1, \ldots, z_{i-1})$ or x_k. Hence,

$$\begin{aligned} x_k &= \Pi(\Pi(z_1, \ldots, z_{i-1}), x_k) \\ &= \Pi(\Pi(z_1, \ldots, z_{i-1}, z_{i+1}, \ldots, z_p), x_k) && \text{by (*)} \\ &= \Pi(z_1, \ldots, z_{i-1}, z_{i+1}, \ldots, z_p, x_k) && \text{by Lemma 9.3.1} \\ &= \Pi(y_1, \ldots, y_m). \ \blacksquare \end{aligned}$$

THEOREM 9.3.1. *Suppose* $\alpha \in S$, β *is a minimum P-undominated subset of* α, *and* $x \in \beta$. *Then for some s,* Π, *and F, s is a nonredundant* α*-sequence,* Π *is a binary, serial MUSTACH operation relative to F,* Π *either is conservative or has a linear tie breaker relative to F, and* $\Pi(s) = x$.

Proof. Let ρ be any linear ordering of A that x bears to every element of $\alpha - \beta$. Let Π be a binary, serial MUSTACH operation relative to F such that either ρ is a tie breaker for Π or else Π is conservative relative to F. Let $(w_1, \ldots, w_r)$ be any nonredundant $\alpha - \beta$-sequence. By hypothesis, β is either the unit set of a P-undominated element of α or else a top P-cycle in α. Either way, there is a β sequence $(x_1, \ldots, x_k)$, $k \geq 1$, for which

$$x = x_k P x_{k-1} P \cdots x_2 P x_1.$$

Let $(y_1, \ldots, y_m)$ be the result of deleting from $(x_1, \ldots, x_k)$ all but the last occurrence of each x_i that occurs more than once. Then $(y_1, \ldots, y_m, w_1, \ldots, w_r)$ is a nonredundant α-sequence. So it suffices to prove that $\Pi(y_1, \ldots, y_m, w_1, \ldots, w_r) = x$.

Because $x_{i+1}Px_i$, $i = 1, 2, \ldots, k-1$, we have

$$\Pi(x_i, x_{i+1}) = x_{i+1}, \qquad i = 1, 2, \ldots, k-1.$$

And because β is a P-undominated subset of α, no w_i bears P to $x = x_k$. Thus, because either Π is conservative or ρ is a tie breaker for Π relative to F,

$$\Pi(x, w_j) = x, \qquad j = 1, 2, \ldots, r.$$

Hence, by Lemma 9.3.2,

$$\Pi(y_1, \ldots, y_m, w_1, \ldots, w_r) = x_k = s. \quad \blacksquare$$

Multistage Choice and GETCHA

If we weaken Theorem 9.3.1 slightly by requiring that Π have a linear tie breaker (rather than that Π *either* have a linear tie breaker *or* be conservative), then we can strengthen this theorem another way: Instead of supposing that x belongs to the GOCHA set from α, we can suppose, more generally, that x belongs to the *GETCHA* set from α—the minimum P-dominant subset of α. So everything choosable according to GETCHA is the result of a binary, serial choice with linear tie breaker and nonredundant agenda sequence. Because the GETCHA set tends to be quite large, multistage choice processes tend to be quite indeterminate. The result of a multistage choice, indeed a binary, serial choice, depends to a striking degree on factors naturally and normally regarded as arbitrary, random, or unpredictable, to wit, the agenda sequence—even though constrained to be nonredundant—and the tie-breaking procedure—even though constrained to reflect a linear tie breaker.

LEMMA 9.3.3. *Suppose* $\alpha \in S$, δ *is the minimum P-dominant subset of* α, and $x \in \delta$. *Then there is a* δ*-sequence* $(x_1, \ldots, x_k)$ *such that* $x_k = x$ *and not* x_iPx_{i+1}, $i = 1, 2, \ldots, k-1$.

Proof. Let $(y_1, \ldots, y_m)$ be a δ-sequence for which $y_m = x$. Then it suffices to show, for $i = 1, 2, \ldots, m-1$, that there are $z_1, \ldots, z_r \in \delta$ such that $z_1 = y_i$, $z_r = y_{i+1}$, and not z_jPz_{j+1}, $j = 1, 2, \ldots, r-1$. Let λ be the least set β containing y_i and containing everything in δ to which something in β does not bear P. Then it suffices to prove that $y_{i+1} \in \lambda$, and for this it suffices to prove that $\lambda = \delta$. By construction, λ is a nonempty subset of $\delta \subseteq \alpha$, and everything in λ bears P to everything in

$\delta - \lambda$, hence to everything in $\alpha - \lambda$. So λ is a P-dominant subset of α. Hence, because δ is the *minimum* P-dominant subset of α, $\lambda \not\subset \delta$. Therefore, $\lambda = \delta$. ■

THEOREM 9.3.2. *Suppose* $\alpha \in S$, δ *is the minimum P-dominant subset of* α, *and* $x \in \delta$. *Then for some s,* Π, and F, *s is a nonredundant* α-*sequence,* Π *is a binary, serial MUSTACH operation with a linear tie breaker relative to F,* and $\Pi(s) = x$.

Proof. By Lemma 9.3.3, there is a δ-sequence $(x_1, \ldots, x_k)$ such that

$$x_k = x$$

and

$$\text{not } x_i P x_{i+1}, \qquad i = 1, 2, \ldots, k-1.$$

Let ρ be any linear ordering of A such that

$$x = x_k \rho x_{k-1} \rho \cdots \rho x_2 \rho x_1.$$

Let Π be a binary, serial MUSTACH operation relative to F for which ρ is a tie breaker relative to F.

Let $(y_1, \ldots, y_m)$ be the result of deleting from $(x_1, \ldots, x_k)$ all but the last occurrence of each x_i that occurs more than once. Let $(w_1, \ldots, w_r)$ be any nonredundant $(\alpha - \delta)$-sequence. Then $(y_1, \ldots, y_m, w_1, \ldots, w_r)$ is a nonredundant α-sequence, and it suffices to prove that $\Pi(y_1, \ldots, y_m, w_1, \ldots, w_r) = x$.

Since δ is a P-dominant subset of α, we have

$$x = x_k P w_j, \qquad j = 1, 2, \ldots, r,$$

whence

$$\Pi(x_k, w_j) = x_k, \qquad j = 1, 2, \ldots, r.$$

But

$$x_{i+1} \rho x_i \text{ and not } x_i P x_{i+1}, \qquad i = 1, 2, \ldots, k-1.$$

Thus

$$\Pi(x_i, x_{i+1}) = x_{i+1}, \qquad i = 1, 2, \ldots, k-1.$$

By Lemma 9.3.2, then,

$$\Pi(x_1, \ldots, x_k, w_1, \ldots, w_r) = x_k = x. \ ■$$

When you make a multistage choice from α, you can end up anywhere in the minimum *P*-dominant subset of α—which could well be the whole of α—even if you are restricted to using a binary, serial division function, a nonredundant agenda sequence, and a tie-breaking procedure representable by a linear ordering.

9.4. A GENERAL DETERMINACY THEOREM

For *binary* multistage choices, this is the extent of indeterminacy. When you make a binary multistage choice of *any* sort from α, you cannot get out of the minimum *P*-dominant subset of α, however much you monkey with the division function, agenda sequence, or tie-breaking procedure:

> If $x_1, \ldots, x_k \in A$ and Π is any binary MUSTACH operation, $\Pi(x_1, \ldots, x_k)$ belongs to the minimum *P*-dominant subset of $\{x_1, \ldots, x_k\}$.

What about *nonbinary* choices? That depends on *C*. If the feasible set β at some stage in a multistage choice from α has a choice set not contained in the minimum *P*-dominant subset of β, then the final choice might be foreign to the minimum *P*-dominant subset of α. Example: *xPy, xPz,* and *yPz*, so that $\{x\}$ is the minimum *P*-dominant subset of $\{x, y, z\}$; yet $C(\{x, y, z\}) = \{z\}$. Π is a MUSTACH operation relative to *F*, and $F(x, y, z) = ((x), (y), (z))$. So $\Pi(x, y, z) = z \neq x$.

On the other hand, so long as the choice *at each stage* in a multistage choice from α belongs to the minimum *P*-dominant subset of the feasible set *at that stage*, the final choice from α must belong to the minimum *P*-dominant subset of α itself. That is:

> If $x_1, \ldots, x_k \in A$ and Π is a GETCHA-consistent MUSTACH operation relative to *F*, then $\Pi(x_1, \ldots, x_k)$ belongs to the minimum *P*-dominant subset of $\{x_1, \ldots, x_k\}$,

where:

> DEFINITION Π *is a* GETCHA-consistent MUSTACH operation relative to *F iff* Π *is a MUSTACH operation relative to F, and for every k and all* $s, s_1, \ldots, s_k \in SEQ$, *if* $F(s) = (s_1, \ldots, s_k)$ *then* $\Pi(s)$ *belongs to the minimum P-dominant subset of* $\{\Pi(s_1), \ldots, \Pi(s_k)\}$.

So a GETCHA-consistent MUSTACH operation represents a multi-stage choice process in which the choice at each stage belongs to the minimum P-dominant subset of the feasible set *at that stage*

So long as $C(\alpha)$ is contained in the minimum P-dominant subset of α for every $\alpha \in S$, *every* MUSTACH operation is perforce GETCHA consistent. And regardless of C, every *binary* MUSTACH operation is GETCHA consistent.

THEOREM 9.4.1. *Every binary MUSTACH operation relative to F is GETCHA consistent relative to F.*

Proof. Let s be a member of SEQ, and let $F(s) = (s_1, s_2)$. It suffices to prove that $\Pi(s)$ belongs to the minimum P-dominant subset of $\{\Pi(s_1), \Pi(s_2)\}$. But

$$\Pi(s) \in C(\{\Pi(s_1), \Pi(s_2)\}).$$

By Axioms C1 and C2 and the definition of P, if $\Pi(s_1)P\Pi(s_2)$ then $\{\Pi(s_1)\}$ is both $C(\{\Pi(s_1), \Pi(x_2)\})$ and the minimum P-dominant subset of $\{\Pi(s_1), \Pi(s_2)\}$; if $\Pi(s_2)P\Pi(s_1)$ then $\{\Pi(s_2)\}$ is both $C(\{\Pi(s_1), \Pi(s_2)\})$ and the minimum P-dominant subset of $\{\Pi(s_1), \Pi(s_2)\}$; finally, if neither $\Pi(s_1)P\Pi(s_2)$ nor $\Pi(s_2)P\Pi(s_1)$ then $\{\Pi(s_1), \Pi(s_2)\}$ is both $C(\{\Pi(s_1), \Pi(s_2)\})$ and the minimum P-dominant subset of $\{\Pi(s_1), \Pi(s_2)\}$. In each case, $\Pi(s)$ belongs to the minimum P-dominant subset of $\{\Pi(s_1), \Pi(s_2)\}$. ■

By virtue of this last theorem, the result about binary MUSTACH operations advertised earlier in this section is a special case of the similar result about GETCHA-consistent MUSTACH operations, to whose proof I now turn.

THEOREM 9.4.2. *Suppose $x_1, \ldots, x_k \in A$ and Π is a GETCHA-consistent MUSTACH operation relative to F. Then $\Pi(x_1, \ldots, x_k)$ belongs to the minimum P-dominant subset of $\{x_1, \ldots, x_k\}$.*

Proof. By induction on k. Let $F(x_1, \ldots, x_k) = (s_1, \ldots, s_r)$. Because Π is GETCHA-consistent relative to F, $\Pi(x_1, \ldots, x_k)$ belongs to the minimum P-dominant subset of $\{\Pi(s_1), \ldots, \Pi(s_r)\}$. For $i = 1, 2, \ldots, r$, let α_i be the set of terms of s_i. So

$$\bigcup_{i=1}^{r} \alpha_i = \{x_1, \ldots, x_k\}.$$

Let δ be the minimum P-dominant subset of $\{x_1, \ldots, x_k\}$; to prove that $\Pi(x_1, \ldots, x_k) \in \delta$. By construction, δ is nonempty. But

$$\delta \subseteq \{x_1, \ldots, x_k\} = \bigcup_{i=1}^{r} \alpha_i.$$

So for some i, $\delta \cap \alpha_i$ is nonempty. Since every member of δ bears P to every member of $\{x_1, \ldots, x_k\} - \delta$, every member of $\delta \cap \alpha_i$ bears P to every member of $\alpha_i - \delta$. Thus, since $\delta \cap \alpha_i$ is nonempty, $\delta \cap \alpha_i$ is a P-dominant subset of α_i. But by inductive hypothesis, $\Pi(s_i)$ belongs to the minimum P-dominant subset of α. So $\Pi(s_i) \in \delta \cap \alpha_i$. Therefore, $\{\Pi(s_1), \ldots, \Pi(s_r)\} \cap \delta$ is nonempty. But since every member of δ bears P to every member of $\{x_1, \ldots, x_k\} - \delta$, every member of $\{\Pi(s_1), \ldots, \Pi(s_r)\} \cap \delta$ bears P to every member of $\{\Pi(s_1), \ldots, \Pi(s_r)\} - \delta$. So $\{\Pi(s_1), \ldots, \Pi(s_r)\} \cap \delta$ is a P-dominant subset of $\{\Pi(s_1), \ldots, (s_r)\}$. Hence, since $\{\Pi(x_1, \ldots, x_k)\}$ belongs to the minimum P-dominant subset of $\{\Pi(s_1), \ldots, \Pi(s_r)\}$, $\Pi(x_1, \ldots, x_k)$ belongs to $\{\Pi(s_1), \ldots, \Pi(x_r)\} \cap \delta$ and thus to δ. ■

COROLLARY 9.4.1. *Suppose $x_1, \ldots, x_k \in A$ and Π is a binary MUSTACH operation relative to F. Then $\Pi(x_1, \ldots, x_k)$ belongs to the minimum P-dominant subset of $\{x_1, \ldots, x_k\}$.*

Proof. Immediate from Theorems 9.4.1 and 9.4.2. ■

COROLLARY 9.4.2. *Suppose $C(\alpha)$ is a subset of the minimum P-dominant subset of α for every $\alpha \in S$. And suppose $x_1, \ldots, x_k \in A$ and Π is a MUSTACH operation. Then $\Pi(x_1, \ldots, x_k)$ belongs to the minimum P-dominant subset of $\{x_1, \ldots, x_k\}$.*

Proof. Immediate from Theorem 9.4.2. ■

Theorems 9.3.2, 9.3.4, 9.4.1, and 9.4.2 are summarized thus:

THEOREM 9.4.3. *If $\alpha \in S$, the following sets are identical:*

(a) *the minimum P-dominant subset of α;*
(b) *$\{x \mid x = \Pi(s)$ for some nonredundant α-sequence s, some binary, serial division function F, and some MUSTACH operation Π for which there is a linear tie breaker relative to $F\}$;*
(c) *$\{x \mid \Pi(s) = x$ for some α-sequence s and some binary MUSTACH operation Π relative to some $F\}$;*
(d) *$\{x \mid \Pi(s) = x$ for some α-sequence s and some GETCHA-consistent MUSTACH operation Π relative to some $F\}$.*

Hence, the following four conditions are equivalent:

GETCHA $C(\alpha) \equiv$ the minimum P-dominant subset of α.

SIMPLE SERIALITY ASSUMPTION (*SSASS*) $C(\alpha) \equiv \{x \in \alpha \mid \Pi(s) = x$ *for some nonredundant* α*-sequence s, some binary, serial division function F, and some MUSTACH operation* Π *for which there is a linear tie breaker relative to F*$\}$.

BINARY ANALYZABILITY (*BINANA*) $C(\alpha) \equiv \{x \in \alpha \mid \Pi(s) = x$ *for some* α*-sequence s and some binary MUSTACH operation* Π *relative to some F*$\}$.

MULTISTAGE CHOICE PROPERTY (*MUSTCHOP*) $C(\alpha) \equiv \{x \in \alpha \mid \Pi(s) = x$ *for some* α*-sequence s and some GETCHA-consistent MUSTACH operation* Π *relative to some F*$\}$.

9.5. SPECIAL CASES

Recall from sction 6.3 that GETCHA is equivalent to GOCHA if, and only if, PI-Transitivity holds (if *xPyIz* then *xPz*). Recall as well that PI-Transitivity is equivalent to PP-Connexity (if *xPyPz* then either *xPz* or *zPx*) plus I-Transitivity. So assuming PI-Transitivity, Theorem 9.4.3 tells us that anything chosen from α by a GETCHA-consistent MUSTACH operation belongs not only to the minimum *P*-dominant subset of α but to the union of minimum *P*-undominated subsets of α—not only to the GETCHA set from α but to the GOCHA set.

THEOREM 9.5.1. *Assuming PI-Transitivity, if* $\alpha \in S$ *then these sets are identical:*

(a) *the union of minimum P-undominated subsets of* α;
(b) *the minimum P-dominant subset of* α;
(c) $\{x \mid \Pi(s) = x$ *for some nonredundant* α*-sequence s, some binary, serial division function F, and some MUSTACH operation* Π *with a linear tie breaker relative to F*$\}$;
(d) $\{x \mid \Pi(s) = x$ *for some nonredundant* α*-sequence s and some binary, serial, conservative MUSTACH operation* Π *relative to some F*$\}$;
(e) $\{x \mid \Pi(s) = x$ *for some* α*-sequence s and some binary MUSTACH operation* Π *relative to some F*$\}$;
(f) $\{x \mid \Pi(s) = x$ *for some* α*-sequence s and some GETCHA-consistent MUSTACH operation* Π *relative to some F*$\}$.

Proof. Immediate from theorems 6.3.3 and 9.4.3. ■

If we allow even a single violation of either PP-Connexity or I-Transitivity (which are conjointly equivalent to PI-Transitivity), we can no longer equate (a) with (b), (c), (e), or (f) in Theorem 9.5.1. For consider a single generic violation of PP-Connexity: *xPyPzIx*. Here $\{x\}$ is the one and only minimum *P*-undominated subset of $\{x, y, z\}$ although $\{x, y, z\}$ itself is the minimum *P*-dominant subset, and $\Pi(z, x, y) = y \notin \{x\}$ if Π is binary and serial and the tie between z and x is broken in z's favor. Or consider a single generic violation of I-Transitivity: *xIyIzPx*. This time $\{z, y\}$ is the union of minimum *P*-undominated subsets of $\{x, y, z\}$, although $\{x, y, z\}$ itself is the minimum *P*-dominant subset, and $\Pi(y, z, x) = x \notin \{y, z\}$ if Π is binary and serial and the tie between y and z is broken in y's favor and the tie between y and x in x's favor.

So if we drop I-Transitivity and assume PP-Connexity by itself, we cannot infer that the result of every multistage choice from a set α belongs to the union of minimum *P*-undominated subsets of α, even if the choice is binary and serial. But we *can* infer this for binary, serial choices that use the *Conservative Rule:*

> Assume PP-Connexity. If $x_1, \ldots, x_k \in A$ and Π is a binary, serial, conservative MUSTACH operation relative to F, then $\Pi(x_1, \ldots, x_k)$ belongs to the union of minimum *P*-undominated subsets of $\{x_1, \ldots, x_k\}$.

What is more, if we replace the assumption of PP-Connexity with the stronger:

> PIP-CONNEXITY *If xPyIzPw then xPw or wPx,*

we can then replace the requirement that Π be binary with the weaker requirement that Π be strongly GOCHA consistent, in this sense:

> DEFINITION Π *is a* strongly GOCHA-consistent MUSTACH operation relative to *F iff* Π *is a MUSTACH operation relative to F, and for every k and all s,* $s_1, \ldots, s_k \in SEQ$, *if* $F(s) = (s_1, \ldots, s_k)$ *then* $C(\{\Pi(s_1), \ldots, \Pi(s_k)\})$ *is the union of minimum P-undominated subsets of* $\{\Pi(s_1), \ldots, \Pi(s_k)\}$.

That is, we can prove the following:

> Assuming PIP-Connexity, if $x_1, \ldots, x_k \in A$ and Π is a serial, conservative, strongly GOCHA-consistent MUSTACH operation

relative to F, then $\Pi(x_1, \ldots, x_k)$ belongs to the union of minimum P-undominated subsets of $\{x_1, \ldots, x_k\}$.

Although stronger than PP-Connexity, PIP-Connexity is weaker than the conjunction of PP-Connexity and I-Transitivity. Just as PP-Connexity is the result of subtracting P-Acyclicity (the suborder condition) from P-Transitivity (the strict-partial-order condition), so PIP-Connexity is the result of subtracting P-Acyclicity from PIP-Transitivity (the interval-order condition): PIP-Connexity and P-Acyclicity are mutually independent and conjointly equivalent to PIP-Transitivity.

I call the property just defined *strong* GOCHA consistency because it is not quite parallel to GETCHA consistency. Where GETCHA consistency requires that the *choice* at each stage *belong to* the GETCHA set from the feasible set at that stage, strong GOCHA consistency requires that the *whole choice set* at each stage *be* the GOCHA set from the feasible set at that stage.

Obviously, all MUSTACH operations are strongly GOCHA consistent so long as C satisfies GOCHA. And binary ones are perforce strongly GOCHA consistent, independently of C.

THEOREM 9.5.2. *Every binary MUSTACH operation relative to F is strongly GOCHA consistent relative to F.*

Proof. Similar to Theorem 9.4.1. ■

To prove the two propositions lately advertised, I need a definition and a lemma.

DEFINITION β *is a* maximum ρ-cycle in α *iff (1)* $\beta \subseteq \alpha \in S$, (2) β *is either a unit set or a* ρ*-cycle, and* (3) $\beta \subset \gamma \subseteq \alpha$ *for no* ρ*-cycle* γ.

Note that a maximum ρ-cycle can be a *unit set*, hence not a cycle at all.

LEMMA 9.5.1. *Assume PP-Connexity. Let* μ *be a maximum P-cycle in a.*

(a) *If* $y \in \alpha - \mu$ *and y bears P to* some *member of* μ, *y bears P to* every *member of* μ.

(b) *If* $\beta \subseteq \alpha$, $\beta \not\subseteq \mu$, β *is a P-cycle, and* some *member of* β *bears P to* some *member of* μ, *then* every *member of* β *bears P to* every *member of* μ.

Proof. (a) Let x be any member of μ; to prove that yPx. Trivial if μ is a unit set. Otherwise, μ is a P-cycle; for some k there are $x_1, \ldots, x_k \in \mu$ such that

$$yPx_1Px_2P \cdots x_{k-1}Px_k = x;$$

and for $i = 1, 2, \ldots, k$, it suffices to show, by induction on i, that yPx_i. yPx_1 by construction. And yPx_{i-1} by inductive hypothesis. Thus, by *PP*-Connexity, since $x_{i-1}Px_i$, either yPx_i or x_iPy. But if x_iPy then $\mu \cup \{y\}$ would be a P-cycle, contrary to the maximum character of μ. Hence, yPx_i.

(b) Let $y \in \beta$ and $x \in \mu$; to prove that yPx. If μ and β overlapped, $\mu \cup \beta$ would be a P-cycle. But then, because μ is a *maximum* P-cycle in α, β would be a subset of μ, contrary to the hypothesis of the lemma. Therefore, μ and β are disjoint, and thus, by virtue of (a), because some member of β bears P to some member of μ, some member of β bears P to x. Hence, because β is a P-cycle, for some k there are $y_1, \ldots, y_k \in \beta$ such that

$$y = y_1Py_2P \cdots Py_k = x.$$

So for $i = 1, 2, \ldots, k$, it suffices to show, by induction on $k - i$, that y_iPy_k. By construction, $y_{k-1}Py_k$. And by inductive hypothesis, $y_{i+1}Py_k$. But y_iPy_{i+1}. By PP-Connexity, then, either y_iPy_k or y_kPy_i. If y_kPy_i, however, $\mu \cup \beta$ would be a P-cycle, contrary to the maximum character of μ. Consequently, y_iPy_k. ■

THEOREM 9.5.3. *Let Π be any serial, conservative, strongly GOCHA-consistent MUSTACH operation relative to F. Suppose that either* (a) *PIP-Connexity holds or else* (b) *PP-Connexity holds and F is binary. Then if $x_1, \ldots, x_r \in A$, $\Pi(x_1, \ldots, x_r)$ belongs to some minimum P-undominated subset of $\{x_1, \ldots, x_r\}$.*

Proof. By induction on r. Since PIP-Connexity implies PP-Connexity, the latter condition holds, by hypothesis of the Theorem, in any case. Since F is serial, there is a $k = 1, 2, \ldots, r - 1$ such that

$$F(x_1, \ldots, x_r) = ((x_1, \ldots, x_k), (x_{k+1}), \ldots, (x_r))$$

Therefore, since Π is a conservative, strongly GOCHA-consistent MUSTACH operation relative to F,

$$\Pi(x_1, \ldots, x_r) \in C(\{\Pi(x_1, \ldots, x_k), x_{k+1}, \ldots, x_r\}),$$

$C(\{\Pi(x_1, \ldots, x_k), x_{k+1}, \ldots, x_r\})$ is the union of minimum P-undominated subsets of $\{\Pi(x_1, \ldots, x_k), x_{k+1}, \ldots, z_r\}$,

and

$$\Pi(x_1, \ldots, x_r) = \Pi(x_1, \ldots, x_k) \text{ iff}$$

$$\Pi(x_1, \ldots, x_k) \in C(\{\Pi(x_1, \ldots, x_k), x_{k+1}, \ldots, x_r\}).$$

Let

$$c_1 = \Pi(x_1, \ldots, x_k),$$

$$c_2 = \Pi(c_1, x_{k+1}, \ldots, x_r),$$

$$\alpha_1 = \{x_1, \ldots, x_k\},$$

and

$$\alpha_2 = \{c_1, x_{k+1}, \ldots, x_r\}.$$

Then there is a β_2 such that

β_2 is a minimum P-undominated subset of α_2, and $c_2 \in \beta_2$.

And by inductive hypothesis, there is a β_1 such that

β_1 is a minimum P-undominated subset of α_1, and $c_1 \in \beta_1$.

Furthermore,

$c_2 = c_1$ iff c_1 belongs to some minimum P-undominated subset of α_2.

Since F is binary only if $r - k = 1$, hence only if α_2 contains nothing more than c_1 and c_2,

(1) PIP-Connexity holds if $\{c_1, c_2\} \neq \alpha_2$.

It suffices to prove that c_2 belongs to some minimum P-undominated subset of $\alpha_1 \cup \alpha_2$. But β_2 is a unit set or P-cycle. Thus, since $\alpha_1 \cup \alpha_2$ is finite, there is a maximum P-cycle μ in $\alpha_1 \cup \alpha_2$ such that

$$\beta_2 \subseteq \mu.$$

Then $c_2 \in \mu \subseteq \alpha_1 \cup \alpha_2$, and μ is a unit set or a P-cycle. So it suffices to show that nothing in $(\alpha_1 \cup \alpha_2) - \mu$ bears P to anything in μ. Suppose, on the contrary, that $y \in (\alpha_1 \cup \alpha_2) - \mu$ and y bore P to something in μ; to deduce a contradiction.

By Lemma 9.5.1(a),

(2) yPv for all $v \in \mu$.

Thus, since $\beta_2 \subseteq \mu$ and β_2 is a minimum *P*-undominated subset of α_2, $y \notin \alpha_2$, whence

(3) $y \in \alpha_1 - \mu$.

β_1 is $\{c_1\}$ or a *P*-cycle. So if β_1 overlapped μ, $\mu \cup \beta_1$ would be a unit set or *P*-cycle, whence, by virtue of the maximum character of μ, β_1 would be a subset of μ, and thus, by (2) and (3), we should have $y \in \alpha_1 - \beta_1$ and $yPc_1 \in \beta_1$, contrary to the assumption that β_1 is a *P*-undominated subset of α_1. Hence, β_1 and μ are disjoint. It follows that

(4) $c_1 \neq c_2$.

It also follows that $c_1 \in \alpha_2 - \beta_2$. By lemma 9.5.1, however, if something in β_1 bore *P* to something in μ, then c_1Pc_2. But since $c_1 \in \alpha_2 - \beta_2$ and β_2 is a *P*-undominated subset of α_2, this is impossible. Hence,

(5) nothing in β_1 bears *P* to anything in μ.

So, by virtue of (2),

$y \notin \beta_1$,

and thus, by (3), since β_1 is a *P*-undominated subset of α_1,

(6) yPw for no $w \in \beta_1$.

If some $w \in \beta_1$ bore *P* to y, then for every $v \in \mu$ we should have $wPyPv$ by (2), and thus, by PP-Connexity, either wPv, contrary to (5), or else vPw, in which case $\mu \cup \{y, w\}$ would be a *P*-cycle, contrary to the maximum character of μ. Therefore, nothing in β_1 bears *P* to y, and so, by virtue of (6),

(7) yIv for all $v \in \beta_1$.

If some $v \in \mu$ bore *P* to some $w \in \beta_1$, then $yPvPw$ by (2), whence, by virtue of PP-Connexity, yPw or wPy, contrary to (7). Hence, nothing in μ bears *P* to anything in β_1, and thus, by (5),

(8) wIv for all $w \in \beta_1$ and $v \in \mu$.

So

(9) c_1Ic_2.

Consequently, since Π is conservative relative to F, if $\alpha_2 = \{c_1, c_2\}$ then $c_2 = c_1$, contrary to (4). Therefore, $\alpha_2 \neq \{c_1, c_2\}$, whence, by (1), PIP-Connexity holds.

By (4), since Π is conservative and strongly GOCHA consistent relative to F, there is a $z \in \alpha_2$ such that

(10) zPc_1.

Thus, by (7) and (2), zPc_1IyPc_2. Consequently, by PIP-Connexity, zPc_2 or c_2Pz. But since $c_2 \in \beta_2$ and β_2 is a P-undominated subset of β_2, if zPc_2 then $z \in \beta_2 \subseteq \mu$, whence c_1Iz by (8), contrary to (10). So c_2Pz. Therefore, by (10) and PP-Connexity, either c_1Pc_2 or c_2Pc_1. But this contradicts (9), completing the proof. ■

The restrictions in this theorem are essential. Weaken any of them the least bit and you make the theorem false, even if you further restrict it to very small, nonredundant agenda sequences. This is demonstrated by the following four examples:

Example 1. P has the form:

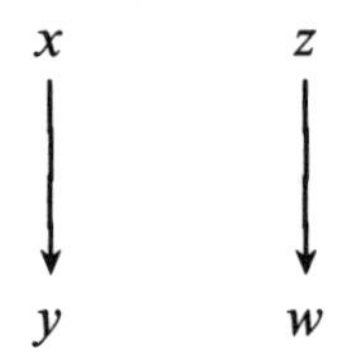

So PIP-Connexity does not hold, although PP-Connexity does. Π is a serial, conservative, strongly GOCHA-consistent MUSTACH operation relative to a nonbinary division function F, which looks like figure 9.9. So $\Pi(z, y) = y$, and $\Pi(y, z, w, x) \in C(\{\Pi(z, y), w, x\}) = \{x, w\}$; say $\Pi(y, z, w, x) = w$. But $\{x, z\}$ is the union of minimum P-undominated subsets

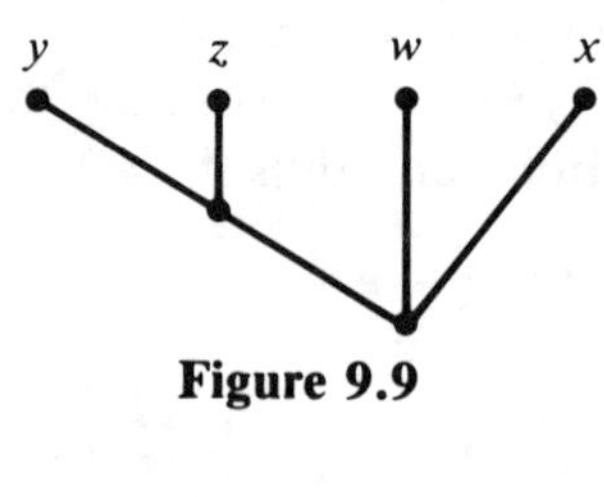

Figure 9.9

of $\{x, y, z, w\}$. This shows we cannot allow the slightest violation of PIP-Connexity in the nonbinary case.

Example 2. P has the form:

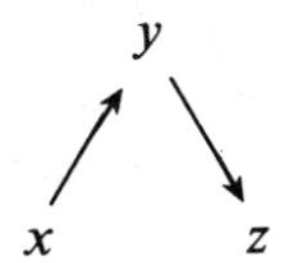

So PP-Connexity (and therewith PIP-Connexity) fails, although I-Transitivity holds. Π is a binary (hence strongly GOCHA-consistent), serial, conservative MUSTACH operation relative to F. So $\Pi(z, x, y) = y$. But $\{x\}$ is the one and only minimum P-undominated subset of $\{x, y, z\}$. This shows we cannot allow even the slightest violation of PP-Connexity in the binary case.

Example 3. P has the form:

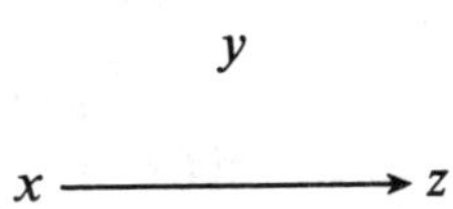

So I-Transitivity fails. But PP-Connexity and even PIP-Connexity hold. Π is a conservative MUSTACH operation relative to a binary but nonserial division function F, which looks like figure 9.10. Thus, $\Pi(z, y, x) = z$. But $\{x, y\}$ is the union of minimum P-undominated subsets of $\{x, y, z\}$. This shows we cannot allow even the slightest violation of the requirement that F be serial.

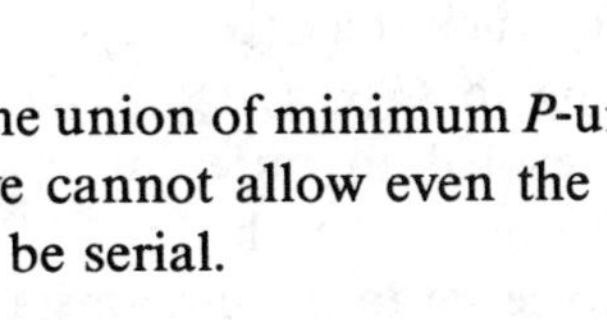

Figure 9.10

Example 4. P has the same form as in the last example. Π is a nonconservative MUSTACH operation relative to a binary, serial division function F. ρ is a linear tie breaker for Π relative to F, and $z\rho y\rho x$. Thus, $\Pi(y, x, z) = z$. But $\{x, y\}$ is the union of minimum P-undominated subsets of $\{x, y, z\}$. This shows we cannot allow even the slightest violation of the requirement that Π be conservative relative to F.

By virtue of Theorems 9.3.2 and 9.5.3, *serial, conservative* choices—*evolutionary* choices, we might call them—are more determinate than other multistage choices.

9.6. CYCLES AND SERIAL CHOICE

Sometimes it is alleged that multistage choice processes, especially serial ones, behave anomalously in the presence of collective-preference cycles, or that collective-preference cycles are manifestly anomalous in serial-choice situations. When there is a P-cycle, the story goes, the result of a serial choice can depend on the agenda sequence, and that is supposed to be a Bad Thing.

The appeal here is to the principle that the result of a serial choice should never depend on the agenda sequence. This means a MUSTACH operation Π relative to a serial division function should satisfy:

Π-SYMMETRY *If* $s \in SEQ$ *and* s' *is a permutation of* s, $\Pi(s') = \Pi(s)$.

Suppose $xPyPzPx$, contrary to P-Acyclicity. Let Π be a binary, serial MUSTACH operation. Then $\Pi(x, y, z) = z$ but $\Pi(z, y, x) = x$, whence $\Pi(x, y, z) \neq \Pi(z, y, z)$, contrary to Π-Symmetry. Because $\{x, y, z\}$ is a P-cycle, two potential agenda sequences, (x, y, z) and (z, y, x), have led to different final choices from the same set.

What of it? According to GOCHA and GETCHA, and quite possibly SOCO 1 and 2 as well (depending on individual preferences and the underlying power structure), $C(\{x, y, z\}) = \{x, y, z\}$. Suppose so. Then z and x both belong to $C(\{x, y, z\})$. What did it matter that different agenda sequences led to different final outcomes? Both outcomes, after all, belong to the choice set from $\{x, y, z\}$—an apparently reasonable choice set in the circumstances.

True, the final outcome depends on the arbitrary selection of an agenda sequence. But since $C(\{x, y, z\})$ has more than one member, the actual choice from $\{x, y, z\}$ must depend on some sort of arbitrary selection anyway.

Besides, Π-Symmetry can fail even when there are no *P*-cycles, indeed even when P + I-Transitivity holds. Example: *P* and *I* are transitive. *xPz, yPz*, and *xIy*. Π is a conservative MUSTACH operation relative to a binary, serial division function. So $\Pi(x, y, z) = x$ and $\Pi(y, x, z) = y$, contrary to Π-Symmetry.

When we choose serially from a set α, different α-sequences will naturally yield different final choices if $C(\alpha)$ has more than one member—a phenomenon that can occur even in the absence of *P*-cycles or other failures of transitivity. It may be desirable that every α-sequence yield a member of $C(\alpha)$. That different α-sequences yield different members merely reflects the multiple membership of $C(\alpha)$. Π-Symmetry says Π must choose the same alternative from α no matter what the agenda sequence. But when $C(\alpha)$ has more than one member, why favor any *one* member of $C(\alpha)$ as the target of *every* serial choice from α? In choosing among the members of $C(\alpha)$, some sort of arbitrary selection must be made anyway. It might as well be the selection of an agenda sequence.

P-cycles are sometimes alleged to cause trouble of another sort. Suppose again that *xPyPzPx*, contrary to P-Acyclicity. Let Π be a MUSTACH operation relative to a binary, serial division function. Then

$$\begin{aligned} \Pi(y, x, z) &= z \\ \Pi(y, x, z, y) &= y \\ \Pi(y, x, z, y, x) &= x \\ \Pi(y, x, z, y, x, z) &= z \end{aligned}$$

and so on without end. Does this not show that a serial choice from a *P*-cycle can always be overturned, giving rise to an unending series of unstable choices?

Not necessarily. Not if there are suitable institutional or other restrictions on the agenda sequence. Certainly not if *redundant* sequences are proscribed. To be sure, one can imagine institutional settings that would permit unending or at least very repetitious series of unstable choices (although it is not obvious that such settings exist in the real world, and it is not obvious that an unending or repetitious series of unstable choices would perforce be a Terrible Thing). What is important is that there are common institutional settings in which a *P*-cycle cannot cause an unending or even very repetitious series of unstable choices. *P*-cycles do not necessarily cause trouble—maybe they never cause trouble—for serial choices.

If serial choice processes behave anomalously, the anomaly has little, if anything, to do with *P*-cycles, and the anomaly is not that serial choices are continually overturned or that different agenda sequences yield different final outcomes. It is rather, I suggest, that different agenda sequences yield *too many* final outcomes, even when the choice process is severely constrained in natural ways, owing to the relatively large size the GETCHA set tends to have. The anomaly, if there is one, is not that the agenda sequence is a determining factor, but that it is too significant a determining factor.

One reason this might be found disturbing is that it means chairmen often have a disturbing degree of power over committee choices. In light of Theorem 9.3.2, a committee chairman can lead his committee to any outcome in the GETCHA set just by the way he breaks ties and arranges the agenda sequence—even if he has no control over how members vote or what gets on the agenda, and even if he is required not to allow repetitions and to break ties as if he were thereby expressing a linear preference.

But I suspect that those most disturbed by large collective-choice sets are those who believe that social institutions for making collective choices are supposed to maximize social welfare or group utility or some such thing. If, instead, one takes my view that the principal justification for such institutions is that they institutionalize such power shifting, or governmental change, as would otherwise occur in a more violent, less predictable, and therefore more costly manner, then one will find nothing puzzling or objectionable about large collective-choice sets. On the contrary, institutions that issue in small choice sets limit the possibilities of successful institutional activity, increasing the incentive to act outside the institutions.

The picture of multistage choice I have been painting is subject to an important qualification. Suppose a multistage choice is to be made from α. Before the multistage process takes place, a coalition g, with the power to dictate the choice of x from α, forms and agrees to support x. At every stage in the process, if the feasible set at that stage contains x, the members of g all support x, even if some members prefer other alternatives feasible at that stage to x. As a result, x is chosen, regardless of the agenda sequence, tie-breaking procedure, or division function. In other words, the multistage choice process itself might be a mere formality, the real choice having been made ahead of time, possibly in one fell swoop. In such a case, the set of alternatives that might be chosen is not the GETCHA set or even the GOCHA set but (as I have, in effect, argued in chapter 7) the SOCO 2 set.

CHAPTER TEN

CHOOSING THE SET FROM WHICH TO CHOOSE

10.1. WHAT IS FEASIBLE?

The distinction between feasible and infeasible alternatives, by whatever name or none, plays a starring role in actual choices and in choice theory, individual and collective. One of the hardest things choosers do, very often, is to decide what the feasible alternatives are—what set to choose from. And one cannot fully describe, explain, or evaluate a choice without somehow citing the set whence it was made—and perhaps saying why *that* set qualifies as the feasible set. As for choice *theory*, choice functions are functions of potential feasible sets, the important independence conditions require independence of infeasible alternatives, the traditional "rationality" conditions require choices from different potential feasible sets to conform to certain patterns, and the peculiarity of multistage choice is that it involves splitting one feasible set into smaller feasible sets and choosing from it by choosing from them.

What makes an alternative feasible? What distinguishes the feasible alternatives on any occasion from the other members of A?

There is no general answer to this question. The criteria of feasibility vary not only from situation to situation but from point of view to point of view. Even with circumstances fixed, there can be conflicting but equally legitimate criteria of feasibility, hence distinct sets with equally legitimate claims to the title of feasible set.

Take the 1936 U.S. presidential election. Any of the following can be regarded, more or less correctly, as the feasible set:

(a) {Roosevelt, Landon};
(b) every candidate officially listed in at least one state;
(c) every major contender for the Republican or Democratic nomination;
(d) everyone whose name was formally placed in nomination or who received at least one vote at the national convention of any political party, major or minor;
(e) every living person who met the Constitutional requirements for President.

Franklin Roosevelt can correctly be said to have been chosen from each of these sets. And the list can be extended indefinitely.

The difference between (a) and (c) is one of situation, of context, of occasion. The numbers of (c) were feasible, one might reasonably contend, early in 1936 and in 1936 as a whole: they alone actively sought the presidency with some significant chance of success during those periods. But by the same test, only the members of (a) were feasible specifically in the Fall. By contrast, (a) and (b) do not differ situationally: both sets can be regarded, from different points of view, as the feasible set on election day. Neither do (c), (d), and (e) differ situationally. Although the feasible set can of course vary from situation to situation, the situation alone does not determine the feasible set.

The feasible alternatives on any occasion might also be called the possible choices, or the alternatives that can be or could have been chosen, on that occasion. But this is not very helpful. Words like "possible" and "can"—and "feasible" itself, for that matter—have elastic meanings. In a very narrow (or strong) sense, only the members of (c) were possible choices in 1936 as a whole: they alone could have been chosen in 1936. In a wider (weaker) sense, each member of (d) and even of (e) was a possible choice. In a wide enough sense, George Washington was a possible choice: one can imagine him returning from the dead.

Possibility of any sort is consistency with some set of *background truths*—the empty set in the case of logical possibility, the laws of physics in the case of physical possibility, the legal requirements for office plus known facts about voter preferences in the case of electoral feasibility, and so on. Expanding the set of background truths contracts the set of possibilities. The trouble is that the set of background truths not only varies from context to context but, depending on one's interests, can comprise any truths whatever.

The difference between feasible and merely conceivable alternatives, between a feasible set and the rest of A, is not that clear. Seeking to make it clearer, we might try characterizing the feasible alternatives on any occasion as those that stand some reasonable chance of being chosen on that occasion by the choice processes at issue. But this blurs the difference between the feasible set and the choice set.

Although criteria of feasibility vary widely, some recur frequently. Here are several, none of them exact or easily made exact:

The feasible alternatives are those the choosers deliberate about. (Just how much attention must one give an alternative to be said to have deliberated about it?)

The feasible alternatives are those whose choice is consistent with all choice-determining factors other than internal motivational (conative) ones. In other words, feasibility is determined by such constraints as choosers' power, authority, resources, information, and opportunities, but not by their preferences, objectives, moral values, or the like. (What about such borderline constraints as coercion, compulsion, moral and legal prohibitions, and mandated budget ceilings?)

The feasible alternatives are those whose choice is consistent with available information. (When, if ever, does information one is not actively entertaining but has learned or should have learned or could easily obtain count as available?)

The feasible alternatives are those the choosers are responsible for choosing or rejecting. (When is who responsible for what?)

The feasible alternatives are those it would not be thoroughly silly or absurd or unreasonable to choose. (What qualifies as silly or absurd or unreasonable?)

The feasible alternatives are those that meet some specified institutional requirement—a candidate-filing law, a legislative agenda-setting rule, or the like. (Institutional requirements are sometimes unclear and sometimes hard to identify, they vary from institution to institution, and to *define* institution-relative feasibility as feasibility according to the institution in question would be circular.)

Although there is no general rule for distinguishing the feasible from the infeasible alternatives on any occasion, there still is a general reason for drawing the distinction. Other things equal, it is easier to choose from a less inclusive set than from a more inclusive set. Hence it is convenient for choosers (individual or collective) to confine their deliberations to relatively small sets, rejecting without deliberation everything outside those sets. Other things equal, it is easier to explain, predict, evaluate, or prescribe a choice from a less inclusive set than from a more inclusive set. Hence it is convenient for him who would explain, predict, evaluate, or prescribe a choice to confine his analysis to a relatively small set, assuming as obvious or as already established that there was or is no basis for expecting or commending the choice of anything outside that set. If you try to explain the 1936 U.S. presidential election as a choice from (e), you must explain, among other things, how (e) got pared to (c). But you do not have to do this if you try to explain the election to begin with as a choice from the less inclusive set (c).

To treat α as the feasible set on any occasion is to ignore (for the time being, anyway) the members of $A - \alpha$. It is to take for granted—to take as obvious or as already established—that there is no reason to choose them, or to expect them to be chosen. It is thereby to limit the scope of one's reasoning about the choice in question. We distinguish feasible from infeasible alternatives precisely in order to draw such a limit. Unable to reason about everything at once, humans must take some things for granted. *The feasible alternatives on a given occasion from a given point of view are those alternatives whose rejection on that occasion is not taken for granted from that point of view.* Although this formulation draws the feasible/infeasible distinction in a completely general way, it makes the distinction relative to a point of view as well as an occasion.

Because we have wide latitude in the way we distinguish feasible from infeasible alternatives, it is important, when examining a specific choice process, to know how sensitive it is to variations in the feasible set. In section 10.2, I formulate several *insensitivity conditions*, capturing several ways a choice function can be insensitive to variations in the feasible set, and I examine the types of choice function that fulfill these conditions.

Choice theorists of all sorts commonly treat feasible sets as raw material given by "nature" and fed into choice processes. But we can also view feasible sets as determined, at least in part, by the choice processes themselves. In section 10.3 and 10.4, I examine three general ways a choice process can choose a feasible set before choosing an alternative from it:

Way 1. By *reduction*—by starting with a large set and winnowing clearly rejectable alternatives.

Way 2. By *adaptation*—by starting with a relatively small set, making a tentative choice from it, and periodically considering new options.

Way 3. By *specification*—by making a relatively inspecific choice (war vs. peace), then progressively making this choice more specific (nuclear vs. conventional war), thereby treating the set of more specific versions of the earlier choice as the new feasible set.

Way 1 is appropriate when the large starting set clearly contains *every* alternative that ought to be considered, although possibly others as well. Way 2 is appropriate when the small starting set clearly contains *only* alternatives that ought to be considered, although possibly not all of them. Adaptation processes come to the fore when we try to solve problems less by deliberation than by innovation—by creating new schemes, new ideas, new technologies. The advantage of Way 3 is that one can start with a set that is both small and unquestionably exhaustive ({war, peace}), even if not as specific as one would like, then make it more specific only when and to the extent that need dictates.

Reduction and adaptation processes transform an initial feasible set by additions or deletions. Specification processes transform an initial feasible set rather by choosing an alternative from it and transforming that alternative into a new feasible set—a set of more specific versions of that alternative.

10.2. INSENSITIVITY TO WHAT IS FEASIBLE

The Conditions: WARP and Variations

The simplest insensitivity condition is also preposterously strong. It says all potential feasible sets have the same choice set:

(I) $C(\alpha) \equiv C(\beta)$.

Given Axioms C1 and C2, (I) implies that A has no more than one member. *Proof:* If $x, y \in A$ then $C(\{x\}) = \{x\}$ and $C(\{y\}) = \{y\}$ by Axioms C1 and C2, but $C(\{x\}) = C(\{y\})$ by (i), so $x = y$.

We may regard an insensitivity condition as postulating some sort of insensitivity to *alterations* in the feasible set, an alteration being an addition or deletion or combination of the two. Another overly strong insensitivity condition, but one that is on the right track, says the choice set is invariant under any alteration in the feasible set:

(II) $C(\alpha) \equiv C(\alpha - \beta)$.

Read from left to right, (II) says that if we eliminate some feasible alternatives, reducing α to $\alpha - \beta$, everything originally choosable (from α) is still choosable (from $\alpha - \beta$). The trouble is that some originally choosable alternatives might have been eliminated; *they* cannot still be choosable.

We can get around this problem by postulating, not that all originally choosable alternatives are still choosable, but that those original choosables that were not eliminated—those that belong to the reduced set—are still choosable:

(III) $C(\alpha) - \beta \equiv C(\alpha - \beta)$.

The left-to-right part of this condition is now plausible.

But the right-to-left part is still too strong. It says that if we add feasible alternatives, expanding $\alpha - \beta$ to α, then every original choosable (from $\alpha - \beta$) is still choosable (from α). But this is implausible when some things in the added set are superior to anything in the original set. Erstwhile choosables can fall from grace when confronted with new options.

We avoid this problem if we restrict (III) to the case in which some things choosable from the expanded set (α) belong to the original set

$(\alpha - \beta)$, so that the new choice set $(C(\alpha))$ does not consist entirely of new alternatives (ones drawn from β). This modification of (III) is the well-known:

WEAK AXIOM OF REVEALED PREFERENCE (WARP) *Suppose* $\alpha, \beta \in S$, $\beta \subseteq \alpha$, and $C(\alpha) \not\subseteq \beta$. Then $C(\alpha - \beta) = C(\alpha) - \beta$.

In section 1.3 I listed six conditions of "rational choice": BICH alone and BICH combined with P-Transitivity (the strict partial-order condition), with PIP-Transitivity (the interval-order condition), with PIP + IPP-Transitivity (the semiorder condition), and with P + I-Transitivity (the weak-order condition). The last is the strongest. It is equivalent to WARP, as Kenneth Arrow proved in 1959.

Several ways of weakening WARP yield other "rationality" conditions. Each of the five "rationality" conditions just cited is equivalent to one or a pair of the following five insensitivity conditions, all of which follow from WARP.

W1 *Suppose* $\alpha, \beta \in S$. *Then* $C(\alpha) \cap C(\beta) = C(\alpha \cup \beta) \cap \alpha \cap \beta$.
W2 *Suppose* $\alpha, \beta \in S$, $\beta \subseteq \alpha$, *and* $C(\alpha) \not\subseteq \beta$. *Then* $C(\alpha - \beta) \subseteq C(\alpha)$.
W3 *Suppose* $\alpha, \beta \in S$, $\beta \subseteq \alpha$, *and* $C(\alpha) \not\subseteq \beta$. *Then if either* $C(\beta) \not\subseteq C(\alpha)$ *or* $C(\alpha - C(\beta)) - \beta \subseteq C(\alpha)$, $C(\alpha - \beta) \subseteq C(\alpha)$.
W4 *Suppose* $\alpha, \beta \in S$, $\beta \subseteq \alpha$, *and* $C(\alpha) \not\subseteq$ b. *Then if* $C(\beta) \not\subseteq C(\alpha)$, $C(\alpha - \beta) \subseteq C(\alpha)$.
W5 *Suppose* $\alpha, \beta \in S$. *Then if* $\beta \subseteq \alpha - C(\alpha)$, $C(\alpha - \beta) \subseteq C(\alpha)$.

Take two sets of alternatives, one a subset of the other. Suppose some choosable members of the bigger set (α) belong to the smaller set $(\alpha - \beta)$. Then they are exactly the choosable members of the smaller set, according to WARP. So read from right to left, WARP says that if we alter the feasible set by *eliminating* some members, then every non-eliminated choosable alternative is still choosable. Read from left to right, WARP says this: If we alter the feasible set by *adding* some members, then every original choosable is still choosable if any is.

W1 says that if we divide the feasible set $(\alpha \cup \beta)$ into two (possibly overlapping) subsets (α, β), then the alternatives (if any) that are choosable from *both* subsets are precisely the common elements of the two subsets that are choosable from the whole set. So read from left to right, W1 says that if we *expand* the feasible set (from α to $\alpha \cup \beta$) by adding another (overlapping) set to it, then every original choosable that is also choosable from the added set is choosable from the expanded set. Read from right to left, W1 says that if we *reduce* the feasible set (to either of two subsets into which it has been divided, say from $\alpha \cup \beta$ to α), then every noneliminated choosable is still choosable.

W2 is just the left-to-right half of WARP. It says that if we expand the feasible set (from $\alpha - \beta$ to α) then everything originally choosable is still choosable if any is.

W3–W5 are qualified versions of W2. They, too, say that if we expand the feasible set (from $\alpha - \beta$ to α) then everything originally choosable (from $\alpha - \beta$) is still choosable (from α) if any is, *provided:*

—in the case of W5, that nothing added (nothing in β) is now choosable (from α);
—in the case of W4, that not everything choosable from the added set (from β) is now choosable (from α);
—in the case of W3, that not everything choosable from the added set is now choosable, unless each original alternative (in $\alpha - \beta$) is now choosable (from α) if choosable from among the original alternatives plus the nonchoosable members of the added set (from $\alpha - \beta$ plus $\beta - c(\beta)$, that is, from $\alpha - C(\beta)$).

W2–W5 can be verbalized also in terms of *reducing* rather than *expanding* the feasible set. According to each, if we eliminate some feasible alternatives (reducing α to $\alpha - \beta$), but not all choosable alternatives (not all of $C(\alpha)$), then nothing originally rejected is now choosable; that is, everything currently choosable (from $\alpha - \beta$) was originally choosable (from β), provided:

—in the case of W5, that nothing choosable (from α) was eliminated (belongs to β);
—in the case of W4, that some things choosable from the eliminated set (from β) were not originally choosable (from α);
—in the case of W3, that some things choosable from the eliminated set were not originally choosable, unless each current alternative (in $\alpha - \beta$) was originally choosable (from α) if choosable from among the current alternatives plus the nonchoosable members of the eliminated set (from $\alpha - \beta$ plus $\beta - C(\beta)$, that is, from $\alpha - C(\beta)$).

Their Logical Relationships

These six insensitivity conditions are logically related as follows:

(i) WARP is stronger than W1; that is, WARP implies but is independent of W1.

Proof that WARP implies W1. Assume WARP; let $\alpha, \beta \in S$; to prove that $C(\alpha) \cap C(\beta) = C(\alpha \cup \beta) \cap \alpha \cap \beta$. If either $C(\alpha \cup \beta) \subseteq \alpha - \beta$ or $C(\alpha \cup \beta) \subseteq \beta - \alpha$, then $C(\alpha) \cap C(\beta)$ and $C(\alpha \cup \beta) \cap \alpha \cap \beta$ are both empty, and the theorem is trivial. Suppose, on the other hand, that

$$C(\alpha \cup \beta) \nsubseteq \alpha - \beta \qquad \text{and} \qquad C(\alpha \cup \beta) \nsubseteq \beta - \alpha.$$

But

$$\alpha \cup \beta \in S \qquad \text{and} \qquad \beta - \alpha \subseteq \alpha \cup \beta.$$

So by WARP, if $\beta - \alpha \in \mathrm{S}$ then

$$C(\alpha \cup \beta) - (\beta - \alpha) = C((\alpha \cup \beta) - (\beta - \alpha)).$$

But

$$C(\alpha \cup \beta) - (\beta - \alpha) = C(\alpha \cup \beta) \cap \alpha$$

and

$$(\alpha \cup \beta) - (\beta - \alpha) = \alpha.$$

Thus, if $\beta - \alpha \in S$ then $C(\alpha \cup \beta) \cap \alpha = C(\alpha)$. On the other hand, if $\beta - \alpha \notin S$ then $\beta - \alpha$ is empty, whence $\alpha \cup \beta = \alpha$, so again $C(\alpha \cup \beta) \cap \alpha = C(\alpha)$. Either way, then,

$$C(\alpha \cup \beta) \cap \alpha = C(\alpha).$$

Similarly,

$$C(\alpha \cup \beta) \cap \beta = C(\beta).$$

Hence,

$$\begin{aligned} C(\alpha) \cap C(\beta) &= C(\alpha \cup \beta) \cap \alpha \cap C(\alpha \cap \beta) \cap \beta \\ &= C(\alpha \cup \beta) \cap \alpha \cap \beta. \ \blacksquare \end{aligned}$$

To prove WARP *independent* of W1, let

(1) $A = \{1, 2, 3\}$, $1P2P3I1$, and $C(A) = \{1\}$.

Then it is routine to check that W1 holds but WARP does not.

(ii) WARP is stronger than W2. It is obvious that WARP implies W2. To prove WARP independent of W2, let

(2) $A = \{1, 2, 3\}$, $1P2P3$, $1P3$, and $C(A) = \{1, 2\}$.

(iii) Besides implying W1 and W2, WARP is implied by them, hence equivalent to their conjunction.

Proof. Assume W1 and W2; let $\alpha, \beta \in S$, $\beta \subseteq \alpha$, and $C(\alpha) \not\subseteq \beta$; to prove that $C(\alpha - \beta) = C(\alpha) - \beta$. By W2, $C(\alpha - \beta) \subseteq C(\alpha)$. But $C(\alpha - \beta) \subseteq \alpha - \beta$. Thus, $C(\alpha - \beta) \subseteq C(\alpha) - \beta$ and it suffices to show that $C(\alpha) - \beta \subseteq C(\alpha - \beta)$. Suppose $x \in C(\alpha) - \beta$; to prove that $x \in C(\alpha - \beta)$. Then

$$x \in \alpha - \beta,$$

and since $\beta \subseteq \alpha$, we have $\alpha = (\alpha - \beta) \cup (\beta \cup \{x\})$, so that

$$x \in C((\alpha - \beta) \cup (\beta \cup \{x\})).$$

But

$$x \in \beta \cup \{x\}.$$

Therefore, $x \in C((\alpha - \beta) \cup (\beta \cup \{x\})) \cap (\alpha - \beta) \cap (\beta \cup \{x\})$. Hence, by W1, $x \in C(\alpha - \beta)$. ■

(iv) W1 and each subsequent axiom are mutually independent. Example (2) proves W1 independent of W2–W5, and example (1) proves W2–W5 independent of W1.

(v) W2 is stronger than W3. It is obvious that W2 implies W3. To prove W2 independent of W3, let

$A = \{1, 2, 3\}$, $1I2I3P1$, and $C(A) = \{2, 3\}$.

(vi) W3 is stronger than W4. It is obvious that W3 implies W4. To prove W3 independent of W4, let

$A = \{1, 2, 3, 4\}$ and $C(\alpha) \equiv \{x \in \alpha \mid x \leq y \text{ for all } y \neq 1 \text{ in } \alpha\}$.

(vii) W4 is stronger than W5. That W4 implies W5 is trivial. To prove W4 independent of W5, let

$A = \{1, 2, 3, 4\}$ and $C(\alpha) \equiv \{x \in \alpha \mid x \neq 2 \text{ if } 1 \in \alpha, \text{ and } x \neq 4 \text{ if } 3 \in \alpha\}$.

To sum up: WARP is stronger than W1, stronger than W2, and equivalent to their conjunction. W1 and each of W2–W5 are mutually independent. And W2 is stronger than W3, which is stronger than W4, which in turn is stronger than W5.

Insensitivity and "Rationality"

I shall prove the equivalence of:

BICH	to W1,
BICH and P-Transitivity	to W1 and W5,
BICH and PIP-Transitivity	to W1 and W4
BICH and PIP + IPP-Transitivity	to W1 and W3, and
BICH and P + I-Transitivity	to W1 and W2,
	hence to WARP

THEOREM 10.2.1. *Assuming Axioms C1 and C2, BICH is equivalent to W1.*

Proof. Assume BICH; let $\alpha, \beta \in S$; to prove that $C(\alpha) \cap C(\beta) = C(\alpha \cup \beta) \cap \alpha \cap \beta$. By virtue of BICH, if $x \in C(\alpha) \cap C(\beta)$ then $x \in \alpha \cap \beta$, yPx for no $y \in \alpha$, and yPx for no $y \in \beta$, whence $x \in \alpha \cup \beta$ and yPx for no $y \in \alpha \cup \beta$, so that $x \in C(\alpha \cup \beta) \cap \alpha \cap \beta$. Conversely, again by virtue of BICH, if $x \in C(\alpha \cup \beta) \cap \alpha \cap \beta$ then $x \in \alpha$, $x \in \beta$, and yPx for no $y \in \alpha \cup \beta$, whence yPx for no $y \in \alpha$ and yPx for no $y \in \beta$, so that $x \in C(\alpha) \cap C(\beta)$.

Now assume W1; let $x \in \alpha \in S$; to prove that $x \in C(\alpha)$ iff, for every $y \in a$, not yPx. If $x \in C(\alpha)$ then for every $y \in \alpha$ we have $\alpha = \alpha \cup \{x, y\}$, so $x \in C(\alpha \cup \{x, y\}) \cap \alpha \cap \{x, y\}$, whence $x \in C(\{x, y\})$ by W1, and thus not yPx by definition of P. Conversely, suppose yPx for no $y \in \alpha$; to prove, by induction on $|\alpha|$, that $x \in C(\alpha)$. Trivial if $\alpha = \{x\}$. Otherwise, let $y \in \alpha - \{x\}$. Then

$$\alpha - \{y\}, \{x, y\} \in S$$

and not yPx, whence

$$x \in C(\{x, y\}).$$

But $\quad x \in C(\alpha - \{y\}) \quad$ by inductive hypothesis.

So

$$x \in C(\{x, y\}) \cap C(\alpha - \{y\}).$$

Hence, by W1, $x \in C((\alpha - \{y\}) \cup \{x, y\}) = C(\alpha)$. ■

LEMMA 10.2.1. *Assume Axioms C1 and C2, BICH, and P-Transitivity, and therewith P-Acyclicity. And suppose $\alpha \in S$ and $x \in \alpha - C(\alpha)$. Then yPx for some $y \in C(\alpha)$.*

Proof. By virtue of BICH, since $x \in \alpha - C(\alpha)$, zPx for some $z \in \alpha$. So for some $m \geq 2$ there are $x_1, \ldots, x_m \in \alpha$ such that

$$x_m = x \quad \text{and} \quad x_i P x_{i+1}, \qquad i = 1, 2, \ldots, m = 1. \tag{*}$$

Were there no largest such m, there would be an m larger than $|\alpha|$ such that (*) held for some $x_1, \ldots, x_m \in \alpha$. But then there would exist i, $j = 1, 2, \ldots, m$ such that $i < j$ and yet $x_i = x_j$, so that

$$x_i P x_{i+1} P \cdots x_{j-1} P x_i,$$

contrary to P-Acyclicity. Consequently, there is a largest m ($m \geq 2$) such that (*) holds for some $x_1, \ldots, x_m \in \alpha$, whence zPx_1 for no $z \in \alpha$, and thus, by virtue of BICH, $x_1 \in C(\alpha)$; and by (*) and P-Transitivity (applied $m - 2$ times), x_1Px. ■

THEOREM 10.2.2. *Assuming Axioms C1 and C2, BICH and P-Transitivity are equivalent to W1 and W5.*

Proof. Assume BICH and *P*-Transitivity. Then W1 holds by Theorem 10.2.1. To deduce W5, suppose $\alpha, \beta \in S$, $\beta \subseteq \alpha - C(\alpha)$, and $x \in C(\alpha - \beta)$; I shall show that $x \in C(\alpha)$. By virtue of BICH, yPx for no $y \in a - \beta$, so it suffices to show that yPx for no $y \in \beta$. By Lemma 10.2.1, since $\beta \subseteq \alpha - C(\alpha)$, if $y \in \beta$ and yPx then zPy for some $z \in C(\alpha)$, whence zPx by P-Transitivity. But since $\beta \subseteq \alpha - C(\alpha)$, $C(\alpha) \subseteq \alpha - \beta$, and thus zPx for no $z \in C(\alpha)$. Hence, yPx for no $y \in \beta$.

Now assume W1 and W5. Then BICH holds by Theorem 10.2.1. To deduce P-Transitivity, suppose $xPyPz$. Then by virtue of BICH, $y \notin C(\{x, y, z\})$ and $z \notin C(\{x, y, z\})$, whence $C(\{x, y, z\}) = \{x\}$. Since xPy, $x \neq y$, so that $\{y\} \subseteq \{x, y, z\} - C(\{x, y, z\})$, and thus, by W5,

$$C(\{x, y, z\} - \{y\}) \subseteq C(\{x, y, z\}) = \{x\}.$$

Since yPz, $y \neq z$. Therefore, since $x \neq y$, $\{x, y, z\} - \{y\} = \{x, z\}$. So

$$C(\{x, z\}) = \{x\}.$$

But since $xPyPz$, $x \neq z$. Hence, by definition of P, xPz.■

THEOREM 10.2.3 *Assuming Axioms C1 and C2, BICH and PIP-Transitivity are equivalent of W1 and W4.*

Proof. Assume BICH and PIP-Transitivity, and therewith P-Transitivity. Then W1 holds by Theorem 10.2.1. To deduce W4, suppose α, $\beta \in S$, $\beta \not\subseteq \alpha$, $C(\alpha) \not\subseteq \beta$, $C(\beta) \not\subseteq C(\alpha)$, and $x \in C(\alpha - \beta)$; I shall show that $x \in C(\alpha)$. By virtue of BICH, since $x \in C(\alpha - \beta)$,

$$yPx \text{ for no } y \in \alpha - \beta, \tag{*}$$

and it suffices to show that yPx for no $y \in \beta$. Suppose, on the contrary, that $y \in \beta$ and yPx; I shall deduce a contradiction. Either $y \in C(\beta)$ or $y \in \beta - C(\beta)$. Even in the latter case, Lemma 10.2.1 implies that there is a $z \in C(\beta)$ such that zPy, whence zPx by P-Transitivity. So in either case there is a z such that

$$z \in C(\beta) \text{ and } zPx.$$

But since $C(\beta) \not\subseteq C(\alpha)$, there is a w such that

$$w \in C(\beta) \text{ but } w \notin C(\alpha).$$

By BICH, since z, $w \in C(\beta)$,

$$wIz.$$

And by Lemma 10.2.1, since $w \notin C(\alpha)$, there is a $v \in C(\alpha)$ such that

$$vPw.$$

By BICH, since $w \in C(\beta)$, $v \notin \beta$, whence $v \in \alpha - \beta$, so by (*), not vPx. But by PIP-Transitivity, since $vPwIzPx$, we have vPx, hence a contradiction.

Now assume W1 and W4. Then BICH holds by Theorem 10.2.1, and since W4 implies W5, P-Transitivity holds by Theorem 10.2.2. To deduce PIP-Transitivity, suppose $xPyIzPw$. Then by P-Transitivity, if zPx, zPy, which is impossible because yIz. Thus, not zPx. But if wPx, then zPx by P-Transitivity. So not wPx. And since xPy, not yPx. So vPx for no $v \in \{x, y, z, w\}$, whence $x \in C(\{x, y, z, w\})$ by BICH. But since $xPyIz$, $x \notin \{y, z\}$. Hence,

$$C(\{x, y, z, w\}) \not\subseteq \{y, z\}.$$

By BICH, since xPy, $y \notin C(\{x, y, z, w\})$. But since yIz, $C(\{y, z\}) = \{y, z\}$. Hence

$C(\{y, z\}) \not\subseteq C(\{x, y, z, w\})$.

Therefore, by W4,

$C(\{x, y, z, w\} - \{y, z\}) \subseteq C(\{x, y, z, w\})$.

Because *yIzPw*, $w \notin \{y, z\}$. But $x \notin \{y, z\}$. So $\{x, y, z, w\} - \{y, z\} = \{x, w\}$. Consequently,

$C(\{x, w\}) \subseteq C(\{x, y, z, w\})$.

By virtue of BICH, since *zPw*, $w \notin C(\{x, y, z, w\})$, whence $w \notin C(\{x, w\})$, so

$C(\{x, w\}) = \{x\}$.

And since *zPw* but not *zPx*, $w \neq x$. Hence, by definition of *P*, *xPw*. ■

THEOREM 10.2.4. *Assuming Axioms C1 and C2, BICH and PIP + IPP-Transitivity are equivalent to W1 and W3.*

Proof. Assume BICH and PIP + IPP-Transitivity, and therewith PIP-Transitivity and P-Transitivity. Then W1 holds by Theorem 10.2.1, and W4 holds by Theorem 10.2.3. To deduce W3, suppose $\alpha, \beta \in S$, $\beta \subseteq \alpha$, $C(\alpha) \not\subseteq \beta$, and either $C(\beta) \not\subseteq C(\alpha)$ or $C(\alpha - C(\beta)) - \beta \subseteq C(\alpha)$. Then if $C(\beta) \not\subseteq C(\alpha)$, $C(\alpha - \beta) \subseteq C(\alpha)$ by W4. Suppose, on the other hand, that $C(\alpha - C(\beta)) - \beta \subseteq C(\alpha)$. Then to show that $C(\alpha - \beta) \subseteq C(\alpha)$, it suffices to show that $C(\alpha - \beta) \subseteq C(\alpha - C(\beta))$. Let $x \in C(\alpha - \beta)$; to prove that $x \in C(\alpha - C(\beta))$. By virtue of BICH,

yPx for no $y \in \alpha - \beta$, (*)

and it suffices to show that *yPx* for no $y \in \beta - C(\beta)$. Let $y \in \beta - C(\beta)$; to prove that not *yPx*. By Lemma 10.2.1, there is a $z \in C(\beta)$ such that

zPy.

Since $C(\alpha) \subseteq \beta$, there is a *w* such that

$w \in C(\alpha) - \beta$.

Then by virtue of BICH, not *zPw*. So either *wPz* or *wIz*. Assuming *wPz*, we have *wPy* by P-Transitivity, and thus if *yPx* then *wPx*. On the other hand, assuming *wIz*, if yPx then *wIzPyPx*, whence *wPx* by virtue of PIP + IPP-Transitivity. So in either case, if *yPx* then *wPx*. But since $w \in C(\alpha) - \beta$, not *wPx* by (*). Hence, not *yPx*.

Now assume W1 and W3. Then BICH holds by Theorem 10.3.1, and since W3 implies W4, PIP-Transitivity holds by Theorem 10.2.3; hence, P-Transitivity holds; and it suffices to prove that, for all x, y, z, w, if $xIyPzPw$ then xPw. Suppose $xIyPzPw$. Then if $x = y$, we have $xPzPw$, whence xPw by P-Transitivity, and the theorem is proved. Suppose, on the other hand, that $x \neq y$. Then it suffices to show that $w \notin C(\{x, w\})$. By BICH, since zPw, $w \notin C(\{x, y, z, w\})$. So it suffices to show that $C(\{x, w\}) \subseteq C(\{x, y, z, w\})$. Since yIx but yPz, $x \neq z$. And since $yPzPw$, we have $w = x$, as well as yPw by P-Transitivity, whence $w \neq y$. So neither x nor w belongs to $\{y, z\}$. Consequently, $\{x, w\} = \{x, y, z, w\} - \{y, z\}$, and it suffices to show that $C(\{x, y, z, w\} - \{y, z\}) \subseteq C(\{x, y, z, w\})$. By W4, then, it suffices to show that

$$C(\{x, y, z, w\}) \not\subseteq \{y, z\} \tag{a}$$

and

$$C(\{x, y, z, w\} - C(\{y, z\})) - \{y, z\} \subseteq C(\{x, y, z, w\}). \tag{b}$$

Since xIy, not yPx. If zPx then $yPzPx$, whence, by P-Transitivity, yPx, which is absurd. So not zPx. If wPx, then $zPwPx$, whence, by P-Transitivity, zPx, which is absurd. So not wPx. Hence, nothing in $\{x, y, z, w\}$ bears P to x, and thus by BICH,

$$x \in C(\{x, y, z, w\}). \tag{**}$$

But $x \neq y$ by assumption. And $x \neq z$ since yPz. Therefore, $x \notin \{y, z\}$. Consequently, by (**), (a) holds, and it suffices to establish (b). Since yPz, $C(\{y, z\}) = \{y\}$. So it suffices to show that

$$C(\{x, y, z, w\} - \{y\}) - \{y, z\} \subseteq C(\{x, y, z, w\}). \tag{b'}$$

Since yPz, $y \neq z$, whence $z \in \{x, y, z, w\} - \{y\}$. But zPw. So by BICH, $w \in C(\{x, y, z, w\} - \{y\}) - \{y, z\}$. But neither y nor z belongs to this set. So at most x belongs, and thus, by (**), (b′) holds. ■

THEOREM 10.2.5. *Assuming Axioms C1 and C2, BICH and P + I-Transitivity are equivalent to W1 and W2, hence to WARP.*

Proof. Assume BICH and P + I-Transitivity. Then W1 holds by Theorem 10.2.1. To deduce W2, suppose $\alpha, \beta \in S$, $\beta \subseteq \alpha$, $C(\alpha) \not\subseteq \beta$, and $x \in C(\alpha - \beta)$; I shall show that $x \in C(\alpha)$. By Lemma 10.2.1, it suffices to show that nothing in $C(\alpha)$ bears P to x. But by virtue of BICH, since

$x \in C(\alpha - \beta)$, nothing in $\alpha - \beta$ bears P to x. Hence, it suffices to show that nothing in $C(\alpha) \cap \beta$ bears P to x. Let $y \in C(\alpha) \cap \beta$; to prove that not yPx. Since $C(\alpha) \not\subseteq \beta$, there exists a $z \in C(\alpha) - \beta$. But by BICH, since $y \in C(\alpha)$, yIz, and since nothing in $\alpha - \beta$ bears P to x, zIx, whence yIx by P + I-Transitivity, and thus not yPx.

Now assume W1 and W2. Then BICH holds by Theorem 10.3.1. And since W2 implies W5, P-Transitivity holds by Theorem 10.3.2. So it suffices to show that I is transitive. Suppose, on the contrary, that I were *not* transitive; then there would exist x, y, z such that $xIyIzPx$, whence I shall deduce a contradiction. Since zPx, $z \neq x$, and since zPx but yIx, $z \neq y$. So z is neither x nor y. But $C(\{x, y, z\}) = \{y, z\}$ by virtue of BICH. So $C(\{x, y, z\}) \not\subseteq \{z\}$, and thus, by W2,

$$C(\{x, y, z\} - \{z\}) \subseteq C(\{x, y, z\}).$$

By BICH, since $x = z$, $x \in C(\{x, y, z\} - \{x\})$. Hence, $x \in C(\{x, y, z\})$. But that is impossible, by virtue of BICH, because zPx. ■

Weaker Conditions

There are innumerable ways to weaken WARP and W1–W5. Here are two examples that merit mention:

W6 *Suppose* $\alpha, \beta \in S$. *Then* $C(\alpha \cup \beta) \cap \alpha \subseteq C(\alpha)$.
W7 *Suppose* $\alpha, \beta \in S$, $\beta \subseteq \alpha$, *and* $C(\alpha) \subseteq \alpha - \beta$. *Then* $C(\alpha) - \beta = C(\alpha)$.

It is fairly obvious, I think, that the first of these insensitivity conditions is equivalent to the left-to-right half of W1. The second is the restriction of WARP to the special case in which the alternatives choosable from the whole (expanded) set all belong to the original (reduced) set.

Although it follows from W1 and therefore from BICH, W6 does not imply BICH. This is proved by letting

$$A = \{1, 2, 3\},\ 1I2P3,\ 1P3, \text{ and } C(A) = \{1\}.$$

But W6 does imply:

STABLE-CHOICE CONDITION *If* $x \in \alpha \in S$ *and* xPy *then* $y \notin C(\alpha)$.

This is the left-to-right half of BICH, the half that implies P-Acyclicity, which I had occasion to discuss in section 6.5. To deduce Stable-

Choice Condition from W6, suppose $x \in \alpha \in S$ and xPy. If, contrary to Stable-Choice Condition, $y \in C(\alpha)$, then $\alpha = \{x, y\} \cup \alpha$ and so $y \in C(\{x, y\} \cup \alpha) \cap \{x, y\}$, whence, by W6, $y \in C(\{x, y\})$, contrary to our assumption that xPy.

Although a consequence of WARP, W7 neither implies nor follows from BICH or P-Acyclicity. For consider the following two interpretations:

$A = \{1, 2, 3\}$, $1P2P3I1$, and $C(A) = \{1\}$.

$A = \{1, 2, 3, 4\}$, $1P2P3P4P1P3$, $4P2$, $C(\{1, 2, 3\}) = \{1\}$, $C(\{1, 2, 4\}) = \{4\}$, $C(\{1, 3, 4\}) = \{1, 3, 4\}$, and $C(\{2, 3, 4\}) = \{2, 3, 4\}$.

In the first interpretation, BICH holds. Hence so does P-Acyclicity. GOCHA holds as well, by the way. But W7 does not hold. In the second, W7 holds, but P-Acyclicity does not. Hence neither does BICH. And neither, as it happens, does GOCHA or GETCHA.

Let me not suggest that W7 has no interesting, transitivity-related consequences. Assume Axioms C1 and C2 and this very modest consequence of BICH:

MINIMUM STABILITY *Suppose* $\alpha \in S$, $|\alpha| = 3$, *and* α *has a P-undominated element. Then* $C(\alpha)$ *contains only P-undominated elements of* α.

Then W7 implies PP-Connexity. I leave the proof to you.

Whatever else may be said about them, BICH and the stronger conditions of "rational choice" enjoy interesting rationales: each is equivalent to an attractive insensitivity condition. *Maximization processes,* broadly construed as choice processes satisfying at least BICH, are especially insensitive to what is feasible—especially robust against perturbations in the feasible set.

This does not mean, however, that no reasonable, interesting insensitivity condition can be satisfied unless BICH holds. A choice function can violate BICH but satisfy W7, after all; and W7 is an especially appealing insensitivity condition. You can, of course, alter the choice set by alterations in the feasible set that add or delete *choosable* alternatives. W7 says that is the *only* way you can alter the choice set: the choice set is invariant under alterations in the feasible set that do not add or delete choosable alternatives.

All the same, insensitivity conditions are surprisingly stringent—and disturbingly so, too, in view of the wide latitude we have in deciding how to distinguish feasible from infeasible alternatives. What this means for the critic or designer of institutions or decision rules, especially ones violating BICH, is that he must evaluate a choice process plus feasibility criterion as a single package. He is free to use any of a number of conflicting criteria of feasibility. Yet the choice process at issue may be highly sensitive to which criterion he uses, yielding acceptable outcomes under one criterion but not under another, equally legitimate criterion. Likewise, the explainer of behavior must treat his feasibility criterion as part of his explanatory model. "Nature" does not tell him which alternatives to treat as feasible. Yet different but equally legitimate feasibility criteria can lead to different predictions even though the hypothesized choice process be the same.

10.3 REDUCTION AND ADAPTATION

A *reduction process* reduces a larger feasible set to a smaller one by eliminating alternatives that are, in some way, clearly rejectable. The larger set may also be called the starting set or the original feasible set. The smaller set may also be called the reduced set or the new feasible set.

By what rule—by what test of "clear rejectability"—is the larger set reduced to the smaller?

There is any number of reasonable-looking reduction rules. Here are five:

RED 1. *The reduced set is the* choice set *from the starting set.*

Interpreted most straightforwardly, this means $C(\alpha)$ is the reduced set when α is the starting set. If the reduction process is continued no further, the chosen alternative will be a member of $C(C(\alpha))$; if it is continued one step further, the chosen alternative will be a member of $C(C(C(\alpha)))$, and so on.

This reduction rule makes a difference—it yields an outcome different from what would be obtained by choosing directly from α—only if $C(C(\alpha)) \neq C(\alpha)$, hence only if C fails to satisfy:

C-CLOSURE $C(C(\alpha)) \equiv C(\alpha)$,

an intuitively attractive and plausibly realistic condition.

To be sure, we might instead interpret the proffered reduction rule to mean there are several choice functions, $C_1, \ldots, C_k$, used to reduce a starting set α to $C_1(\alpha)$, then to $C_2(C_1(\alpha))$, and so on. Example: $n = 3$, α is the set of all living persons, C_1 represents the constitutional requirements for the U.S. presidency, C_2 the combined major-party nominating processes, and C_3 the final presidential-election process. Another example: $n = 2$, α is the set of all possible bundles of Congressional legislation, and C_1 represents a process for setting a budget ceiling.

RED 2. *The reduced set is any subset of the starting set that contains* some *choosable element of the starting set.*

To represent a choice process compatible with this reduction rule, C must fulfill:

W2 *Suppose* $\alpha, \beta \in S$, $\beta \subseteq \alpha$, *and* $C(\alpha) \not\subseteq \beta$. *Then* $C(\alpha - \beta) \subseteq C(\alpha)$.

This means that if a starting set (α) is reduced to a subset ($\alpha - \beta$) containing some choosable element of the starting set, then anything chosen from the reduced set ($\alpha - \beta$) is choosable from the starting set.

W2 is a strong condition. By Theorem 10.2.5, it implies P + I-Transitivity in the presence of BICH. Much the same reasoning shows that W2 implies PI-Transitivity (hence PP-Connexity and I-Transitivity) in the presence of Minimum Stability. I leave the details to you.

RED 3. *The reduced set is* any subset *of the starting set that contains all the choosable members of the starting set.*

To represent a choice process compatible with this rule, C must fulfill the following condition:

Suppose $\alpha, \beta \in S$, $\beta \subseteq \alpha$, and $C(\alpha) \subseteq \alpha - \beta$. Then $C(\alpha - \beta) \subseteq C(\alpha)$.

The condition says that if a starting set (α) is reduced to a subset ($\alpha - \beta$) containing all the choosable elements of the starting set ($C(\alpha) \subseteq \alpha - \beta$), then everything choosable from the reduced set was choosable from the starting set, too ($C(\alpha - \beta) \subseteq C(\alpha)$).

This is just a slight reformulation of W5. And by Theorem 10.2.2, W5 implies P-Transitivity in the presence of BICH. You should have no difficulty adapting the proof of Theorem 10.2.2 to show that W5 implies PP-Connexity in the presence of Minimum Stability.

RED 4. *The reduced set is any P-undominated subset of the starting set.*

To represent a choice process compatible with this rule, C must fulfill:

If $\alpha \in S$ and β is a P-undominated subset of α then $C(\beta) \subseteq C(\alpha)$.

Obviously, any choice function satisfying BICH or GOCHA satisfies this condition.

RED 5. *The reduced set is any P-dominant subset of the starting set.*

To represent a choice process compatible with this rule, C must fulfill:

If $\alpha \in S$ and β is a P-dominant subset of α then $C(\beta) \subseteq C(\alpha)$.

Obviously, any choice function satisfying BICH, GOCHA, or GETCHA satisfies this condition.

In an *adaptation process*, a tentative choice is made from a relatively small starting set, then reconsidered in the light of new options. The process of tentative choice and reconsideration can continue any number of times. At each successive time, a tentative choice is made from a set containing the previous tentative choice and some new alternatives. If—but, of course, only if—the previous tentative choice is still choosable, it remains the tentative choice.

Clearly, an adaptation process is just a multistage choice process of the serial, conservative variety (see ch. 9). And of these we know a good deal: If α comprises all the alternatives considered in a process of this sort (the starting set plus all new alternatives considered along the way), then the alternatives that might be chosen all belong to the GETCHA set from α; they all belong to the GOCHA set if PI-Transitivity holds or if PIP-Connexity holds and the process is strongly GOCHA consistent or if PP-Connexity holds and each set along the way has just two elements; and anything in the GOCHA set can be thus chosen, regardless of C, hence regardless of P.

10.4. CHOICE BY SPECIFICATION

In a *specification process*, a choice is made, individually or collectively, from among relatively inspecific alternatives, then the chosen alternative is made more specific. This means that a set of more specific versions of the chosen alternative—a set of *specifications* of that alternative, let us say—is treated as the new feasible set. The alternative chosen from that set might in turn be made more specific still.

Example 1. In a dispute with Liechtenstein, the Captains Regent of the Republic of San Marino collectively choose war over peace, then nuclear over conventional war, next devastation by dirigible-delivered cobalt bomb over other general types of nuclear war, and so on.

Example 2. A committee of banquet planners chooses luncheon over breakfast, brunch, tea, dinner, or late supper; then al fresco buffet over indoor buffet, indoor sit-down, or box lunch; next vegetarian over fish or meat, and so on.

Formalization

In discussing choice by specification, I regard alternatives as elements not of A but of S—as subsets of A rather than elements of A. This way alternatives can vary in specificity. A relatively large set in S is a relatively inspecific alternative. A unit set is maximally specific. If $\alpha \in S$, a *specification* of α is just a subset of α. In Example 1, *war* is the set of belligerent elements of A and *peace* the set of pacific elements; the two specifications of war, namely *nuclear war* and *conventional war*, are disjoint subsets of war; and *devastation by dirigible-delivered cobalt bomb* is a subset of nuclear war.

Let S^* be the family of potential feasible sets.

DEFINITION $S^* = \{H \subseteq S \mid H$ *is finite and nonempty, and any two members of H are disjoint*$\}$.

Mutual disjointness is needed to ensure mutual exclusiveness (see sections 1.1, 1.2).

Let C^* be a choice function defined on S^* rather than S. Then:

AXIOM C*1 $C^* : S^* \to S^*$.
AXIOM C*2 $C^*(H) \subseteq H$ *for all* $H \in S^*$.

Let P^* be the preference relation corresponding to C^*.

DEFINITION $\alpha P^*\beta$ *iff* $\{\alpha, \beta\} \in S^*$, $\alpha \neq \beta$, *and* $C^*(\{\alpha, \beta\}) = \{\alpha\}$.

On occasion I shall tacitly use the following consequences of Axioms C*1 and C*2 and the definitions of S^* and P^*:

If $H \in S^*$, H is a partition of some member of S.
If $H \in S^*$, $C(H)$ is nonempty.
If $\alpha \in S$, $C^*(\{\alpha\}) = \{\alpha\}$.
If α, β are disjoint members of S and not $\alpha P^*\beta$ then $\beta \in C^*(\{\alpha, \beta\})$.
If $\alpha \in C^*(\{\alpha, \beta\})$, not $\beta P^*\alpha$.
If $\alpha P^*\beta$ then α and β are disjoint members of S.
If $\alpha P^*\beta$, not $\beta P^*\alpha$.
Not $\alpha P^*\alpha$.

A potential feasible set H is *more specific* than another, B, if H is obtainable from B by decomposing some relatively inspecific alternatives found in B (war) into specifications thereof (nuclear war, conventional war).

DEFINITION *H is* more specific *than B iff H, $B \in S^*$, $\bigcup H = \bigcup B$, $|H| > |B|$, and for every $\beta \in B$, some partition of β is a subset of H.*

Thus, {nuclear war, conventional war, peace} is more specific than {war, peace}.

Suppose C^* represents a process of choice by specification. Then the following seem to be reasonable assumptions:

AXIOM C*3 *If H is more specific than B and $\beta \in C^*(B)$, some element of $C^*(\{\gamma \in H | \gamma \subseteq \beta\})$ belongs to $C^*(H)$.*

AXIOM C*4 *If H is more specific than B and $\delta \in B - C^*(B)$, some element of $C^*(\{\gamma \in H | \gamma \subseteq \delta\})$ does* not *belong to $C^*(H)$.*

AXIOM C*5 $C^*(C^*(H)) \equiv C^*(H)$.

AXIOM C*6 *Suppose $H \in S^*$, H has exactly three members, and H has a P^*-undominated member. Then $C^*(H)$ is the set of P^*-undominated members of H.*

Because C^* represents a process of choice by specification, if we choose from the set H by making a relatively nonspecific choice, say the choice of β from the less specific set B, then making this choice more specific, the result must belong to $C^*(H)$. We make β more specific by choosing among the various specifications of β found in H, that is, by choosing from $\{\gamma \in H \mid \gamma \subseteq \beta\}$. So the final choice from H will be an element—it could be any element—of $C^*(\{\gamma \in H \mid \gamma \subseteq \beta\})$. Hence, every element of $C^*(\{\gamma \in H \mid \gamma \subseteq \beta\})$ must belong to $C^*(H)$. Axiom C*3 makes the more modest assertion that *some* element of $C^*(\{\gamma \in H \mid \gamma \subseteq \beta\})$ belongs to $C^*(H)$.

On the other hand, if in choosing from the less specific B we *reject* an alternative δ, so that $\delta \in B - C^*(B)$, then we could *not* necessarily have chosen from H by choosing δ from B and then making δ more specific. So we could not necessarily have chosen from H by choosing among the specifications of δ found in H, that is, by choosing from $\{\gamma \in H \mid \gamma \subseteq \delta\}$. Hence, some element of $C^*(\{\gamma \in H \mid \gamma \subseteq \delta\})$ does *not* belong to $C^*(H)$, which is precisely what Axiom C*4 says.

Axiom C*5 is just C-Closure (see section 10.3) restated for C^* instead of C. I assume $C^*(H)$ always is the result of narrowing the choice from H as far as possible by the process represented by C^*. Axiom C*5 says the choice from H can indeed be narrowed no further.

Axiom C*6 postulates a very weak connection between choices from pair sets and choices from multimember sets. It is far weaker than:

BICH* $C^*(H) \equiv \{\alpha \in H \mid \beta P^* \alpha \textit{ for no } \beta \in H\}$.

What makes BICH* unacceptable is that some H in S^* might have no P^*-undominated element. In other words, given Axioms C*1 and C*2, BICH* implies that P^* is acyclic, and you know what a tough condition that is. But Axiom C*6 postulates BICH* only for the special case in which H does have a P^*-undominated element. What is more, Axiom C*6 further restricts BICH* to three-element feasible sets. And a three-element set with a P^*-undominated member can contain no P^*-cycle. So Axiom C*6 postulates BICH* just for very small, uncomplicated, cycle-free sets. Axiom C*6 follows, by the way, not just from BICH* but from the "*"-ed analog of GOCHA—and of SOCO 1 and 2 as well, assuming "*"-ed versions of DP-Condition and Binary Power.

The GOCHA* Set as Target

What can we say in general about choice sets under a process of choice by specification?

Granting that such a process must satisfy Axioms C*1–C*6, we can say this:

GOCHA* $C^*(H) \equiv$ *the union of minimum* P^**-undominated subsets of* H.

That is, GOCHA* follows from Axioms C^*1–C^*6.

To prove this, I need five lemmata.

LEMMA 10.4.1. *Assume Axioms* C^*1–C^*6. *If* H *and* B *are disjoint members of* S^* *and* $\bigcup HP^*\bigcup B$, *something in* H *bears* P^* *to something in* B.

Proof. By induction on $|H \cup B|$.

Case 1. There are α, β such that $H = \{\alpha\}$ and $B = \{\beta\}$. Then $\alpha = \bigcup HP^* \bigcup B = \beta$ by hypothesis.

Case 2. H has two or more members. Let α_1 be one member. Let $\alpha_2 = \bigcup (H - \{\alpha_1\})$. Then $\bigcup H = \alpha_1 \cup \alpha_2$, so $\alpha_2 \cup \alpha_2 P^* \bigcup B$, and thus $\bigcup B \notin C^*(\{\alpha_1 \cup \alpha_2, \bigcup B\})$. Therefore, by Axiom C^*4, since $\{\alpha_1, \alpha_2, \bigcup B\}$ is more specific than $\{\alpha_1 \cup \alpha_2, \bigcup B\}$, some member of $C^*(\{\bigcup B\})$ does not belong to $C^*(\{\alpha_1, \alpha_2, \bigcup B\})$. That is, $\bigcup B \notin C^*(\{\alpha_1, \alpha_2, \bigcup B\})$. So either $\alpha_1 P^*\bigcup B$ or $\alpha_2 P^*\bigcup B$. But by inductive hypothesis, if $\alpha_1 P^*\bigcup B$ then α_1 bears P^* to something in B, and if $\alpha_2 = \bigcup (A - \{\alpha_1\}) P^*\bigcup B$ then something in $H - \{\alpha_1\}$ bears P^* to something in B. Either way, something in H bears P^* to something in B.

Case 3. B has two or more members. Let $\beta_1 \in B$ and $\beta_2 = \bigcup (B - \{\beta_1\})$. Then $\bigcup B = \beta_1 \cup \beta_2$, so $\bigcup HP^*\beta_1 \cup \beta_2$, and thus $\beta_1 \cup \beta_2 \in C^*(\{\bigcup H, \beta_1 \cup \beta_2\})$. Therefore, by Axiom C*4,

some member of C^* $(\{\beta_1, \beta_2\})$ does not belong to $C^*(\{\bigcup H, \beta_1, \beta_2\})$. (*)

Hence, by Axiom C*6, either β_1 or β_2 is not a P^*-undominated element of $\{\bigcup H, \beta_1, \beta_2\}$.

Suppose first that β_1 is not a P^*-undominated element of $\{\bigcup H, \beta_1, \beta_2\}$. Then either $\bigcup HP^*\beta_1$ or $\beta_2 P^*\beta_1$. But if $\bigcup HP^*\beta_1$, then by inductive hypothesis, some member of H bears P^* to $\beta_1 \in B$, and the Lemma is proved. Assume, on the other hand, that $\beta_2 P^*\beta_1$. Then $C^*(\{\beta_1, \beta_2\}) = \{\beta_2\}$, whence, by (*), $\beta_2 \notin C^*(\{\bigcup H, \beta_1, \beta_2\})$, and thus, by Axiom C*6, β_2 is not a P^*-undominated element of $\{\bigcup H, \beta_1, \beta_2\}$. That is, either $\bigcup HP^*\beta_2$ or $\beta_1 P^*\beta_2$. But since $\beta_2 P^*\beta_1$, not $\beta_1 P^*\beta_2$. Therefore, $\bigcup HP^*\beta_2 = \bigcup (B - \{\beta_1\})$. Hence, by inductive hypothesis, something in H bears P^* to something in $B - \{\beta_1\} \subseteq B$, as was to be proved.

By exactly the same reasoning, if it is β_2 (as opposed to β_1) that is not a P^*-undominated element of $\{\bigcup H, \beta_1, \beta_2\}$, then either some member of H bears P^* to $\beta_2 \in B$, which proves the Lemma, or else $\bigcup HP^*\beta_1$, whence, by inductive hypothesis, something in H bears P^* to $\beta_1 \in B$, which again proves the Lemma. ■

LEMMA 10.4.2. *Assume Axioms C*1–C*6. Suppose* $H \in S^*$, $\beta \in S - H$, β *bears* P^* *to nothing in* H, *and an element* α *of* H *bears* P^* *to* β. *Then* $\bigcup HP^*\beta$.

Proof. If $H = \{\alpha\}$ then $\alpha = \bigcup HP^*\beta$ by hypothesis. Suppose, on the other hand, that H contains more than α. Let $\alpha' = \bigcup (H - \{\alpha\})$. Then $\bigcup H = \alpha \cup \alpha'$. By Lemma 10.4.1, if $\beta P^*\alpha'$, β would bear P^* to something in $H - \{\alpha\}$, contrary to hypothesis. So not $\beta P^*\alpha'$. Thus, since $\alpha P^*\beta$, α is a P^*-undominated element of $\{\alpha, \alpha', \beta\}$ and β is not. Hence, by Axiom C*6, $\beta \notin C^*(\{\alpha, \alpha', \beta\})$. But by Axiom C*3, if $\beta \in C^*(\{\alpha \cup \alpha', \beta\})$ then some member of $C^*(\{\beta\})$ would belong to $C^*(\{\alpha, \alpha', \beta\})$, that is, β would belong to $C^*(\{\alpha, \alpha', \beta\})$, which we have just seen to be impossible. Consequently, $\beta \notin C^*(\{\alpha \cup \alpha', \beta\})$. So $\alpha \cup \alpha' = \bigcup HP^*\beta$. ■

LEMMA 10.4.3 *If* $H \in S^*$ *and* H *is a three-member* P^**-cycle*, $C^*(H) = H$.

Proof. Suppose $C^*(H) \neq H$; to deduce a contradiction. Then $C^*(H)$ has either one or two members.

Suppose $C^*(H)$ has one member, say α. Then there are $\beta, \gamma \in H - C^*(H)$ such that $\alpha P^*\beta P^*\gamma P^*\alpha$. By Axiom C*3, if $\beta \in C^*(\{\alpha \cup \gamma, \beta\})$ then some member of $C^*(\{\beta\})$ belongs to $C^*(\{\alpha, \gamma, \beta\})$; that is, $\beta \in C^*(\{\alpha, \gamma, \beta\})$. Therefore, $\beta \notin C^*(\{\alpha \cup \gamma, \beta\})$, whence $\alpha \cup \gamma \in C^*(\{\alpha \cup \gamma, \beta\})$, and thus, by Axiom C*3, some member of $C^*(\{\alpha, \gamma\})$ belongs to $C^*(\{\alpha, \gamma, \beta\})$. But $\gamma P^*\alpha$, so $C^*(\{\alpha,\gamma\}) = \{\gamma\}$, and thus $\gamma \in C^*(\alpha, \gamma, \beta\}) = \{\alpha\}$, which again is impossible.

Suppose, on the other hand, that $C^*(H)$ has two members. Then there are α, β such that $C^*(H) = \{\alpha, \beta\}$ and $\alpha P^*\beta$, whence

$$C^*(C^*(H)) = C^*(\{\alpha, \beta\}) = \{\alpha\} \neq C^*(H) = \{\alpha, \beta\},$$

contrary to Axiom C*5. ■

LEMMA 10.4.4. *Assume Axioms C*1–C*6. If* $C^*(H)$ *is* B *or more specific than* B, $C^*(B) = B$.

Proof. $C^*(C^*(H)) = C^*(H)$ by Axiom C*5. So if $C^*(H) = B$ then $C^*(B) = B$. If, on the other hand, $C^*(H)$ is more specific than B, we again have $C^*(B) = B$, else there would be an α such that $\alpha \in B - C^*(B)$, whence, by Axiom C*4, some member of $C(\{\gamma \in C^*(H) | \gamma \subseteq \alpha\})$ and hence of $C^*(H)$ would not belong to $C^*(C^*(H))$, contrary to Axiom C*5. ■

LEMMA 10.4.5. *Assume Axioms C*1–C*6. For $H \in S^*$, nothing in $H - C^*(H)$ bears P^* to anything in H.*

Proof. Suppose $\alpha \in H - C^*(H)$, $\beta \in C^*(H)$, and $\alpha P^* \beta$; to deduce a contradiction. Let

$$\gamma = \bigcup (C^*(H) - \{\beta\})$$

so that

$$\bigcup C^*(H) = \beta \cup \gamma.$$

Then if γ is nonempty, $\{\beta, \gamma\} \in S^*$ and $C^*(H)$ is either $\{\beta, \gamma\}$ or more specific than $\{\beta, \gamma\}$, whence $C^*(\{\beta, \gamma\}) = \{\beta, \gamma\}$ by Lemma 10.4.4, so not $\beta P^* \gamma$. But $\alpha P^* \beta$. Therefore, β bears P^* to nothing in $\{\alpha, \gamma\}$, whereas something in $\{\alpha, \gamma\}$ bears P^* to β, and thus, by Lemma 10.4.2, $\alpha \cup \gamma P^* \beta$. If, on the other hand, γ is empty, then $\alpha = \alpha \cup \gamma$, whence $\alpha \cup \gamma P^* \beta$ by hypothesis. Either way, then,

$$\alpha \cup \gamma P^* \beta.$$

Let δ be the union of all members of $H - C^*(H)$ other than α, so that

$$\bigcup (H - C^*(H)) = \alpha \cup \delta.$$

Suppose $\delta P^* \alpha \cup \gamma$; whence I shall deduce a contradiction. Then because $\alpha P^* \beta$, either δ is the sole P^*-undominated element of $\{\alpha \cup \gamma, \beta, \delta\}$, or else $\{\alpha \cup \gamma, \beta, \delta\}$ is a P^*-cycle. By Axiom C*6 and Lemma 10.4.3, then, $C^*(\{\alpha \cup \gamma, \beta, \delta\})$ is $\{\delta\}$ or $\{\alpha \cup \gamma, \beta, \delta\}$. Either way, $\delta \in C^*(\{\alpha \cup \gamma, \beta, \delta\})$. Therefore, by Axiom C*4, since H is more specific than $\{\alpha \cup \gamma, \beta, \delta\}$ and $(H - C^*(H)) - \{\alpha\} = \{\eta \in H\} | \eta \subseteq \delta\}$, some member of $C^*((H - C^*(H)) - \{\alpha\})$ and hence of $H - C^*(H)$ belongs to $C^*(H)$, which is absurd. Consequently,

$$\text{not } \delta P^* \alpha \cup \gamma.$$

Let H' be $\{\alpha \cup \gamma, \beta\}$ or $\{\alpha \cup \gamma, \beta, \delta\}$ according as δ is or is not empty. Then $H' \in S^*$, and $\alpha \cup \gamma$ is a P^*-undominated member of H' but β is not, whence $\beta \notin C(H')$ by Axiom C*6. But H is more specific than H', and $\beta \in H$. So by Axiom C*4, some element of $C^*(\{\beta\})$ does not belong to $C^*(H)$. That is, $\beta \notin C^*(H)$, which is absurd. ■

THEOREM 10.4.1. *Assume Axioms C*1–C*6. For $H \in S^*$, $C^*(H)$ = the union of minimum P^*-undominated subsets of H.*

Proof. Let U = the union of minimum P^*-undominated subsets of H; to prove that $C^*(H) = U$.

Let B be a minimum P^*-undominated subset of H. Then B is not empty. If $B = H$, some member of B belongs to $C^*(H)$. Suppose, on the other hand, that $B \subset H$. Then $H - B$ and B are disjoint members of S^*, and nothing in $H - B$ bears P^* to anything in B, whence not $\bigcup(H - B)P^* \bigcup B$ by Lemma 10.4.1, so that

$$\bigcup B \in C^*(\{\bigcup(H - B), \bigcup B\}). \qquad (*)$$

But H is either $\{\bigcup(H - B), \bigcup B\}$ or more specific than $\{\bigcup(H - B), \bigcup B\}$. If the former then $\bigcup B \in C^*(H)$ by (*), and $B = \{\bigcup B\}$, whence some member of B belongs to $C^*(H)$. And by Axiom C*3, if H is more specific than $\{\bigcup(H - B), \bigcup B\}$, then some member of $C^*(H)$ belongs to $C^*(H)$. But B is either a unit set or a P^*-cycle. If the former then $B \subseteq C^*(H)$. And if B were a P^*-cycle but $B \not\subseteq C^*(H)$ then something in $B - C^*(H)$ would bear P^* to something in $B \cap C^*(H)$, contrary to Lemma 10.4.5. So $B \subseteq C^*(H)$. Hence, since our choice of B was arbitrary, $U \subseteq C^*(H)$. And it suffices to prove that $C^*(H) \subseteq U$, which means that $C^*(H) - U$ is empty.

Suppose $C^*(H) - U$ were *not* empty; to deduce a contradiction. By Lemma 10.4.5, nothing in $H - C^*(H)$ bears P^* to anything in $C^*(H) - U$. If, in addition, nothing in U bore P^* to anything in $C^*(H) - U$, then $C^*(H) - U$ would be a P^*-undominated subset of H, whence some subset of $C^*(H) - U$ would be a minimum P^*-undominated subset of H and, therefore, a nonempty subset of U, which is absurd. Hence, something in U bears P^* to some α in $C^*(H) - U$. But by construction, nothing in $H - U$ bears P^*, hence α does not bear P^*, to anything in U. Therefore, by Lemma 10.4.2,

$$\bigcup U P^* \alpha.$$

Let β be the union of sets other than α that belong to $C^*(H) - U$, so that $\bigcup(C^*(H) - U) = \alpha \cup \beta$. By Lemma 10.4.1, since nothing in $H - U$, hence nothing in $(C^*(H) - U) - \{\alpha\}$, bears P^* to anything in U, not $\alpha P^* \bigcup U$.

Let H' be $\{\bigcup U, \alpha\}$ or $\{\bigcup U, \alpha, \beta\}$ according as β is empty or not. Then $H' \in S$, $C^*(H)$ is H' or more specific than H', and $\bigcup U$ is a P^*-undominated member of H' but α is not, so that $\alpha \notin C^*(H')$ by Axiom C*6, and thus $C^*(H) \neq H'$, contrary to Lemma 10.4.4. ■

This theorem shows that any member of the GOCHA set can be the outcome of a process of choice by specification, depending on the way maximally specific alternatives are grouped to form less specific ones. Consequently, although the GOCHA set tends to encompass less than the GETCHA set, specification processes are somewhat indeterministic and not necessarily Pareto efficient. It is worth noting that this result does not depend on any particular definition of P^* in terms of P.

CHAPTER ELEVEN

MULTIPLE ISSUES

11.1. MULTI-ISSUE CHOICES AND THEIR MULTIPLICITY

Often the final outcome of a collective-choice process is a bundle of positions on as many issues, each issue decided separately. In legislative and committee voting and in referendum elections, the outcome is a bundle of yes-or-no positions on issues decided by separate votes. In decentralized collective-choice processes, such as markets, different individuals or subgroups decide different issues (ones of property transfer, for example), and the final outcome is a position bundle whose choice is attributable to all the individuals or subgroups involved but to no one of them.

Sometimes it is illuminating to look at alternatives as position bundles even though the relevant issues are not decided separately. Example: To vote for a political party or candidate is, in effect, to vote for several issue positions as a single package. The issues in question need not be policy issues in the usual sense. They can include such personal-attribute dimensions as hair color, regional accent, and the number of times the candidate was expelled from Harvard for cheating. Analyzing alternatives into their issue dimensions might help explain the actions of voters and politicians.

Looking at collective-choice alternatives as issue bundles enables us to talk about one of the most prevalent of social phenomena. When individual or corporate creatures with partly conflicting preferences cooperate or compromise, form an alliance or make a deal, they *exchange support across issues:* they support a package of positions, one on each issue involved in the exchange; although none of them favors every position in the package, they agree (in effect) to all the positions because they prefer the package as a whole to the alternative package that would prevail without the exchange. Vote traders obviously exchange support across issues. So do coalition-government partners; for them the issues concern legislation, portfolio assignments, or both. So, implicitly, do the supporters of a political candidate or legislative leader who builds a winning platform from planks that have too little support to be enacted singly. The participants in an ordinary economic trade also exchange support across issues: if I trade you a banana for a coconut, I have supported your favorite position on the issue of who gets the banana in return for your support of my favorite position on the issue of who gets the coconut. The exchange of support across issues—*generalized exchange*—characterizes all political, economic, and other interpersonal activity involving partial conflict of interest.

For our investigations of multi-issue collective choice, let there be m *issues,* $A_1, \ldots, A_m$, each a set—finite or infinite—of alternative positions. If A_i is to be decided before A_j, and if the choice from A_j depends on how A_i is decided, then $i < j$. To the extent that issue indices are not determined by this rule, they are purely arbitrary.

Like the alternatives in A, the positions constituting each A_i are mutually exclusive. But unlike alternatives drawn from disjoint potential feasible sets, positions drawn from different issues are not necessarily mutually exclusive: if x and y belong to disjoint subsets of A, it is not possible to choose both x and y; but if z and w belong to different issues, it may be possible to choose both z and w.

I shall henceforth assume that every alternative in A is an ordered m-tuple $(x_1, x_2, \ldots, x_m)$ for which $x_1 \in A_1$, $x_2 \in A_2$, and so forth. This means that A is a subset of the cartesian product of $A_1, \ldots, A_m$:

AXIOM A $\quad A \subseteq \bigtimes_{i=1}^{m} A_i.$

I do not assume the converse of Axiom A—that every vector of positions, one on each issue, belongs to A. Reason: Sometimes I interpret A as the set of feasible alternatives and each A_i as the set of feasible positions on the ith issue. Then the converse of Axiom A says that every combination of feasible positions on the m issues is itself feasible. As a general condition, that is preposterous. Among other things, it rules out budget constraints: singly affordable (feasible) positions on different issues need not be jointly affordable. Sometimes

I interpret A as the set of *possible* (not necessarily feasible) alternatives. But depending on the exact meaning of "possible," it could be that singly possible positions on different issues are not jointly possible.

Every alternative in A is a possible *total outcome* of the collective-choice process under consideration—total because it comprises positions on *all m* issues. A *partial outcome* is a possible combination of positions on just the first k issues for some $k \leq m$. Let PA be the set of partial outcomes.

DEFINITION $PA = \{(x_1, \ldots, x_k) \mid$ *either* $k = m$ *and* $(x_1, \ldots, x_k) \in A$, *or else* $0 \leq k < m$ *and* $(x_1, \ldots, x_k, x_{k+1}, \ldots, x_m) \in A$ *for some* $x_{k+1}, \ldots, x_m\}$.

Note that every total outcome also is a partial outcome: $A \subseteq PA$. Note also that the *empty vector*, $\emptyset$, is a partial outcome: an element $(x_1, \ldots, x_k)$ of PA is $\emptyset$ if $k = 0$.

11.2. THE UNIVERSAL INSTABILITY THEOREM

Virtually every collective-choice outcome issues in part from a generalized exchange of some sort—from a bargain, compromise, vote trade, treaty, purchase, or the like. From very mild assumptions I shall deduce that any total outcome for which generalized exchange is essential—any total outcome obtainable only through some generalized exchange—must be *unstable* in this sense: some group has the power to block or overturn the given outcome in favor of another outcome that its members like more.

This universal instability theorem has myriad applications. It helps explain the continual shifting of successful legislative alliances, the tendency of ruling parties eventually to lose elections, the instability of coalition governments, and the fugacity of dominant international alliances (compared with the greater durability of dominant single actors, such as Rome). It also shows that the instabilities whose possibility is demonstrated by Condorcet's Voting Paradox example and by various impossibility theorems constitute the rule rather than the exception: far from being peculiar to special institutions, preference profiles, or feasible sets, they afflict almost all political, economic, and other interpersonal activity almost all the time.

Formally, the theorem is about the positive integer m, the sets $A_1, \ldots, A_m$ and A, and a relation Δ and function Γ.

A is now the set of *feasible* total outcomes, and each A_i is the set of feasible positions on the ith issue.

Δ is the relation of a total outcome x to a total outcome y when some group with the power to move from y to x (to block y in favor of x) prefers x to y. So interpreted, $\Delta = D_1^A$ (see section 7.2). The reason I have not defined Δ as D_1^A is that my theorem does not depend on this definition and my assumptions may be plausible for other interpretations of Δ.

The collective-choice process under consideration determines a choice set from each A_i and from A. The positions on A_i that do not belong to the choice set from A_i are the *rejected* positions. Assuming that the first $k - 1$ issues have already been decided, the choice sets from A_k and from A may depend on which positions were in fact chosen from $A_1, \ldots, A_{k-1}$. These choice sets also may depend on whether there is to be any generalized exchange across $A_k, \ldots, A_m$.

When $(x_1, \ldots, x_k) \in PA$, $\Gamma(x_1, \ldots, x_k)$ is the choice set from A, given that $x_1, \ldots, x_k$ have been chosen from $A_1, \ldots, A_k$ and that there is to be no generalized exchange across $A_{k+1}, \ldots, A_m$; in short, $\Gamma(x_1, \ldots, x_k)$ is the choice set from A in the presence of $(x_1, \ldots, x_k)$ and the absence of subsequent exchange. With Γ so intepreted, $\Gamma(x_1, \ldots, x_k)$ is a subset of A, every member of $\Gamma(x_1, \ldots, x_k)$ has the form $(x_1, \ldots, x_k, x_{k+1}, \ldots, x_m)$, and $\Gamma(x_1, \ldots, x_k)$ is simply the no-exchange choice set from A when $k = 0$—when $(x_1, \ldots, x_k) = \emptyset$, that is.

If $(x_1, \ldots, x_{k-1}) \in PA - A$, let $r(x_1, \ldots, x_{k-1})$ be the set of rejected positions on A_k in the presence of $(x_1, \ldots, x_{k-1})$ and the absence of subsequent exchange.

DEFINITION *r is the function on $PA - A$ defined by the condition: $r(x_1, \ldots, x_{k-1}) = A_k - \{x \mid x$ is the kth coordinate of some vector in $\Gamma(x_1, \ldots, x_{k-1})\}$.*

The reason r is defined on $PA - A$ rather than PA as a whole is that if $x \in A$ then $r(x)$ would have to be a subset of the $(k + 1)$th issue, which does not exist. Note that if $(x_1, \ldots, x_k) \in PA - A$ then $\Gamma(x_1, \ldots, x_k)$ is a subset of A but $r(x_1, \ldots, x_k)$ is rather a subset of A_{k+1}.

My theorem rests on four assumptions. One is Axiom A, introduced in section 11.1. Here are two more:

AXIOM $\Gamma 1$ *$x \in \Gamma(x)$ whenever $x \in A$.*

AXIOM Γ2 *If* x_{k+1} *is the* $(k+1)$*th coordinate of some vector in* $\Gamma(x_1, \ldots, x_k)$ *then* $\Gamma(x_1, \ldots, x_k, x_{k=1}) \subseteq \Gamma(x_1, \ldots, x_k)$.

These three conditions follow from my interpretation of A and Γ: they are necessarily true—tautologous, for those who like to use that word that way. They do not restrict the underlying collective-choice process or situation in any way.

The connection between choice sets and the Δ relation is, of course, controversial. My fourth assumption is a very modest one about this connection:

ΔΓ-CONDITION *If* $x_k \in r(x_1, \ldots, x_{k-1})$ *and* $x \in \Gamma(x_1, \ldots, x_{k-1}, x_k)$ *then* $y \Delta x$ *for some* y.

Suppose the first $k-1$ issues have already been decided, and a position $x_k \in A_k$ would be rejected in the absence of any subsequent exchange across $A_k, \ldots, A_m$. Surely, then, some group with the power to reject x_k prefers that it be rejected, except possibly as part of a generalized-exchange package. As captured by ΔΓ-Condition, this means that if a total outcome x contains x_k plus whatever has already been chosen ($x_1, \ldots, x_{k-1}$, say) and is not based on any exchange across $A_k, \ldots, A_m$ (if $x \in \Gamma(x_1, \ldots, x_{k-1}, x_k)$, that is), then x is dispreferred to some feasible outcome y by some group with the power to block x in favor of y.

Example: Three two-sided issues are to be decided by majority voting. So Δ is the relation of majority preference on A, and every generalized exchange is, in effect, a vote trade. x_1 has been chosen from A_1—never mind how or why. In the absence of subsequent vote trading, x_2 would be chosen from A_2 and y_2 rejected. In the presence of x_2 and the absence of subsequent vote trading, x_3 would be chosen from A_3 and y_3 rejected, whereas in the presence of y_2 and the absence of subsequent vote trading, y_3 would be chosen and x_3 rejected. Formally:

$$\Gamma(x_1) = \{(x_1, x_2, x_3)\},$$

so that

$$\Gamma(x_1, x_2) = \{(x_1, x_2, x_3)\},\ r(x_1) = \{y_2\}, \text{ and } r(x_1, x_2) = \{y_3\}.$$

But

$$\Gamma(x_1, y_2) = \{(x_1, y_2, y_3)\}, \text{ so that } r(x_1, y_2) = \{x_3\}.$$

The situation is depicted by a decision tree, in which trade-free choices are circled (figure 11.1). According to ΔΓ-Condition, because $y_2 \in r(x_1)$ and $(x_1, y_2, y_3) \in \Gamma(x_1, y_2)$, some outcome dominates (x_1, y_2, y_3). This means that a majority prefer some outcome, presumably (x_1, x_2, x_3), to (x_1, y_2, y_3).

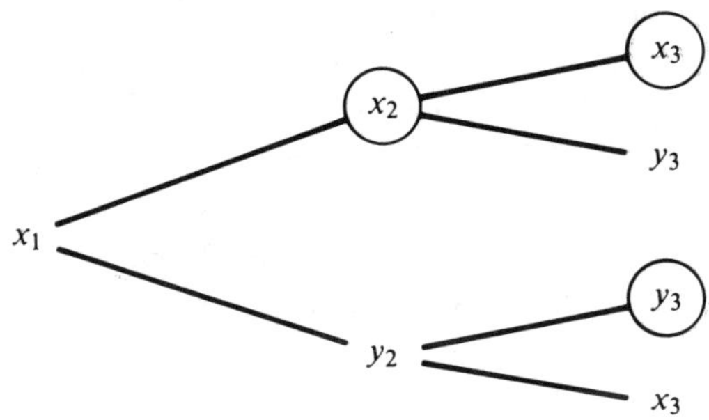

Figure 11.1

Since y_2 would be defeated in the presence of x_1 and the absence of subsequent vote trading, some majority—*g*, let us say—prefer x_2 to y_2 in the presence of x_1. So long as each member of *g* realizes that x_2 would lead to x_3 and y_2 to y_3 in the absence of vote trading, *g* will prefer (x_1, x_2, x_3) to (x_1, y_2, y_3), as required.

But what if those who support x_2 do so only because they mistakenly believe that x_2 will lead to y_3 and y_2 to x_3? Then, it seems, no majority will prefer (x_1, x_2, x_3) to (x_1, y_2, y_3) after all. Thus, ΔΓ-Condition apparently presupposes, for every issue, that sufficiently many actors have either accurate beliefs about choices on subsequent issues, or preferences that do not depend on subsequent choices, or some combination of the two.

I say "apparently," because whether ΔΓ-Condition really is thus restrictive is partly a *semantic* question. Suppose that, in the presence of x_1, a majority prefer x_2 to y_2 only because they mistakenly believe x_2 will lead to y_3 and y_2 to x_3. Although they do not prefer (x_1, x_2, x_3) to (x_1, y_2, y_3) *under the descriptions* "(x_1, x_2, x_3)" and "(x_1, y_2, y_3)," they do prefer (x_1, x_2, x_3) to (x_1, y_2, y_3) under a *different* pair of descriptions, to wit:

(d_1) "the outcome that would prevail if x_1 and x_2 were chosen and there were no vote trading involving A_3"

(d_2) "the outcome that would prevail if x_1 and y_2 were chosen and there were no vote trading involving A_3"

The majority in question fail to realize that "(x_1, x_2, x_3)" and (d_1) describe the same outcome; similarly for "(x_1, y_2, y_3)" and (d_2). It is common for someone to prefer one thing to another under one pair of descriptions ("the 1945 Château Lafite" and "the Vinegro Vineyards Dago Red") but not under another pair of descriptions ("the poisoned wine" and "the unpoisoned wine"), failing to realize that both pairs of descriptions describe the same pair of things (that the Lafite is poisoned whereas the Vinegro Red is not).

In the case at hand, ΔΓ-Condition requires that a majority prefer some outcome, presumably (x_1, x_2, x_3), to (x_1, y_2, y_3). If this means that a majority prefer (x_1, x_2, x_3) to (x_1, y_2, y_3) under *some* pair of descriptions *or other*—restricted, perhaps, to a specified range that includes (d_1) and (d_2) as well as "(x_1, x_2, x_3)" and "(x_1, y_2, y_3)"—then ΔΓ-Condition is satisfied even in the troublesome case lately cited. But if it means that a majority prefer (x_1, x_2, x_3) to (x_1, y_2, y_3) under the specific descriptions "(x_1, x_2, x_3)" and "(x_1, y_2, y_3)," then ΔΓ-Condition might not be satisfied. I favor the former, more liberal interpretation because a preference for an outcome x to an outcome y under any of a range of descriptions that includes (d_1) and (d_2) is a sufficient incentive to block or overturn y in favor of x.

But whichever way one interprets ΔΓ-Condition, this assumption is quite mild as general choice-set constraints go. And it is the only of my assumptions that is in any way restrictive: Axioms A, Γ1 and Γ2 are utterly devoid of empirical content.

Assuming Axiom A, Axioms Γ1 and Γ2, and ΔΓ-Condition, I shall prove that every feasible total outcome foreign to $\Gamma(\emptyset)$, the no-exchange choice set from A, is unstable—dominated under Δ. That, of course, includes every feasible total outcome for which generalized exchange is essential.

THEOREM 11.2.1. *Assuming Axiom A, Axioms Γ1 and Γ2, and ΔΓ-Condition, if $x \in A - \Gamma(\emptyset)$ then $y \Delta x$ for some y.*

Proof. $x \in \times_{i=1}^{m} A_i$ by Axiom A; say $x = (x_1, \ldots, x_m)$. But $x \in \Gamma(x)$ by Axiom Γ1, whereas $x \notin \Gamma(\emptyset)$ by hypothesis. That is, $x \in \Gamma(x_1, \ldots, x_i)$ when $i = m$ but not when $i = 0$. For some $k = 1, 2, \ldots, m - 1$, then,

$$x \in \Gamma(x_1, \ldots, x_k, x_{k+1}) \text{ but } x \notin \Gamma(x_1, \ldots, x_k),$$

whence $$\Gamma(x_1, \ldots, x_k, x_{k+1}) \not\subseteq \Gamma(x_1, \ldots, x_k),$$

and thus, by Axiom Γ2, x_{k+1} is not the $(k + 1)$th coordinate of any vector in $\Gamma(x_1, \ldots, x_k)$. By Axiom A, however, $x_{k+1} \in A_{k+1}$. Therefore,

$$x_{k+1} \in r(x_1, \ldots, x_k).$$

But $\quad x \in \Gamma(x_1, \ldots, x_k, x_{k+1})$.

So $\quad y\Delta x$ for some y $\qquad$ by ΔΓ-Condition. ■

11.3. EIGHT THINGS WORTH NOTING ABOUT THIS THEOREM

Thing 1 Theorem 11.2.1 does *not* say that the outcomes in $\Gamma(\emptyset)$, the no-exchange choice set, are stable, only that no *other* feasible outcomes are stable. The members of $\Gamma(\emptyset)$, some or all of them, may or may not be stable; witness two examples:

Example A Using Simple Majority Rule, three voters are to decide two issues: passage vs. defeat of measure a and passage vs. defeat of measure b. Four outcomes are feasible: both measures pass (ab), a passes but b is defeated ($a\bar{b}$), a is defeated but b passes ($\bar{a}b$), both are defeated ($\overline{ab}$). The voters rank the outcomes in order of preference as follows:

Mr. 1	Mr. 2	Mr. 3
$a\bar{b}$	$\bar{a}b$	$\overline{ab}$
ab	ab	$\bar{a}b$
$\overline{ab}$	$\overline{ab}$	$a\bar{b}$
$\bar{a}b$	$a\bar{b}$	ab

In the absence of vote trading, both measures would be defeated: $\Gamma(\emptyset) = \{\overline{ab}\}$. *POW* is majoritarian, so Δ is the relation of majority preference, and the E_1 relationships within A are the following:

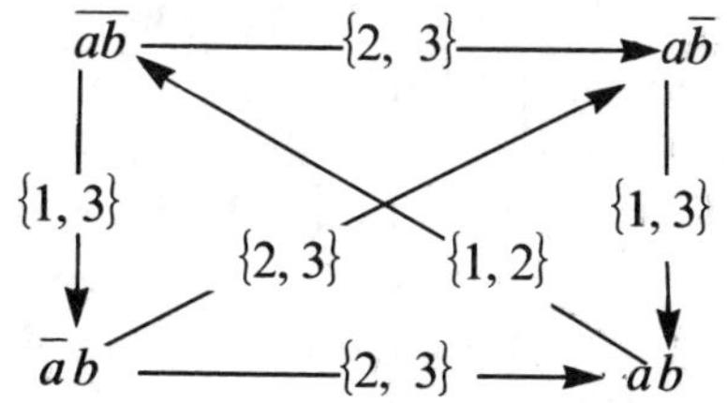

Every feasible outcome, including $\overline{ab}$, is unstable.

Example B Five voters rank the same four outcomes this way:

Mr. 1	Mr. 2	Mr. 3	Mr. 4	Mr. 5
$a\bar{b}$	$a\bar{b}$	$\bar{a}b$	$\bar{a}b$	$\overline{ab}$
ab	$\overline{ab}$	ab	$\overline{ab}$	$a\bar{b}$
$\overline{ab}$	ab	$\overline{ab}$	ab	$\bar{a}b$
$\bar{a}b$	$\bar{a}b$	$a\bar{b}$	$a\bar{b}$	ab

Once again, issues are decided by Simple Majority Rule, *POW* is majoritarian, and $\Gamma(\emptyset) = \{\overline{ab}\}$. But this time the four outcomes are related as follows:

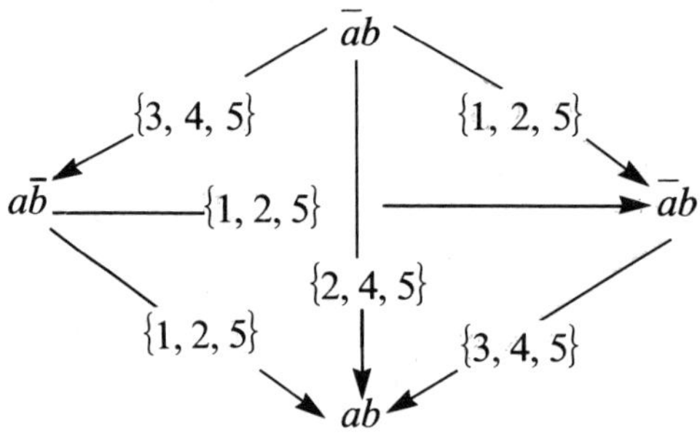

$\overline{ab}$ is now stable.

Thing 2 The theorem does *not* say that every feasible outcome is unstable in the presence of generalized exchange, or even that generalized exchange always *produces* an unstable outcome. Not only can a generalized exchange coexist with a stable outcome, but the very outcome produced by a generalized exchange can be stable (in which case it must belong to $\Gamma(\emptyset)$). The theorem just says every outcome for which generalized exchange is *essential*—every outcome that can *only* result from a generalized exchange—is unstable.

In example B, $\overline{ab}$ (the sole member of $\Gamma(\emptyset)$) is stable. Yet Messrs. 1 and 3 might trade votes, Mr. 1 agreeing to vote for *b* in return for Mr. 3's agreement to vote for *a*. Without further trading, the resulting outcome would be *ab*, which Messrs. 1 and 3 both prefer to $\overline{ab}$. To be sure, Messrs. 2 and 4 could block the effect of this trade by trading votes themselves, Mr. 2 agreeing to vote against *a* in return for Mr. 4's agreement to vote against *b*, ensuring that the actual outcome will be $\overline{ab}$, which Messrs. 2 and 4 both prefer to *ab*—as does Mr. 5. Either way, perfectly rational vote trading (hence generalized exchange) would coexist with the stable $\overline{ab}$. And if the second trade occurred, the very result of trading would be $\overline{ab}$. This is consistent with the theorem because vote trading is not *essential* to $\overline{ab}$: $\overline{ab}$ also is the outcome that would prevail in the absence of any vote trading.

Some might feel that because the two trades cancel each other out, there is no reason to trade. But that cannot be so. If there were no reason for Messrs. 2 and 4 to trade, there would be an excellent reason for Messrs. 1 and 3 to trade, in which case there would be an excellent reason for Messrs. 2 and 4 to trade after all. If there were no reason for Messrs. 1 and 3 to trade, there would be no reason for Messrs. 2 and 4 to trade, either, in which case there would be an excellent reason for Messrs. 1 and 3 to trade after all.

If Messrs. 1 and 3 trade votes but Messrs. 2 and 4 do not, then an *organized minority* (Messrs. 1 and 3) will have frustrated the wishes of a *disorganized majority* (Messrs. 2, 4, and 5)—a majority who, for some reason, could not afford the cost of cooperating. It is perhaps an awareness of this type of situation that explains the complaints we hear about democratic governments serving "special interests" rather than the "public interest."

Thing 3 The theorem does not apply only to voting and vote trading. It also applies, for example, to economic exchange, as illustrated by the following example.

Example C Mr. 1 can exchange a banana for Mr. 2's coconut. Four outcomes are feasible: Mr. 1 gives Mr. 2 a banana and gets a coconut (bc), Mr. 1 gives Mr. 2 a banana but gets no coconut ($b\overline{c}$), Mr. 1 gives no banana but still gets a coconut ($\overline{b}c$), neither gives anything ($\overline{bc}$). Messrs. 1 and 2 rank these outcomes as follows:

Mr. 1	Mr. 2
$\overline{b}c$	$b\overline{c}$
bc	bc
$\overline{bc}$	$\overline{bc}$
$b\overline{c}$	$\overline{b}c$

If no exchange occurred, the resulting outcome would be $\overline{bc}$. That is,

$\Gamma(\emptyset) = \{\overline{bc}\}$. Because *bc* and $\bar{b}c$ differ only in Mr. 1's action, he has the power to block *bc* in favor of $\bar{b}c$, which he prefers to *bc*, so $E_1(\{1\}, \bar{b}c, bc, A)$ (and, therefore, $\bar{b}c \Delta bc$). For the same reason, $E_1(\{1\}, \overline{bc}, b\bar{c}, A)$. And because $b\bar{c}$ and *bc* differ only in Mr. 2's action, as do $\overline{bc}$ and $\bar{b}c$, we likewise have $E_1(\{2\}, b\bar{c}, bc, A)$ and $E_1(\{2\}, \overline{bc}, \bar{b}c, A)$. But because Messrs. 1 and 2 both prefer *bc* to $\overline{bc}$ and together have the power to trade, $E_1(\{1, 2\}, \overline{bc}, bc, A)$. So the four outcomes are E_1-related as follows:

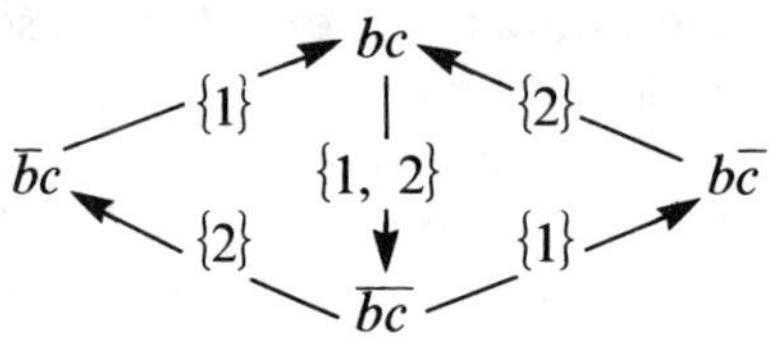

No outcome is stable.

This does *not* mean there will be *no exchange*. An *agreement* to exchange the banana for the coconut is likely to be *enforceable*—by conscience if not coercion. And we might regard the enforcement of such an agreement as the cancellation of Mr. 1's power or incentive to overturn *bc* in favor of $\bar{b}c$ and Mr. 2's power or incentive to overturn *bc* in favor of $b\bar{c}$: *bc* would become stable—but only *after* an enforceable agreement has been made. The dominance structure depicted above, in which *bc* is unstable, represents the *preagreement* situation, in which every enforceable obligation is either a standing obligation or the creature of an earlier agreement.

An enforceable exchange agreement always alters the rules of the game so as to stabilize an initially unstable outcome. Economic exchange is no different in this respect from majority voting: once some majority successfully coalesce, *enforceably agreeing* to support an initially unstable outcome, that outcome can no longer be overturned.

Thing 4 The theorem shows that the instabilities exhibited by the classical voting paradox and similar examples are not mere oddities—that they are not peculiar to certain rare choice processes, feasible sets, or preference profiles. On the contrary, *every* collective-choice outcome to which some form of generalized exchange is essential—and that includes virtually every interesting, realistic collective-choice outcome—is perforce unstable.

Some might contend that elections with small numbers of candidates often have stable outcomes. But that depends on how the feasible alternatives are described. An election that has a stable outcome when the feasible alternatives are taken to be the *candidates* might lack a stable outcome when the feasible alternatives are taken to be the *candidate-platform pairs*, each consisting of a candidate plus a platform that he either did adopt or could have adopted. Looked at this way, even a two-candidate election can have an unstable outcome: the loser might have won had he adopted a different platform.

Thing 5 The theorem helps explain why it is hard to secure cooperation: Because cooperators with partly conflicting preferences exchange support across issues, if cooperation by actors with partly conflicting preferences be needed to achieve a particular outcome then some actors must have the joint power to block or overturn that outcome in favor of another, which they like more. This is illustrated by the celebrated Prisoners' Dilemma:

Example D Two prisoners, Messrs. 1 and 2, must each decide whether to confess to a joint crime. The four feasible outcomes and their consequences are displayed in this payoff matrix:

		Mr. 2	
		does not confess	confesses
Mr. 1	does not confess	Mild penalty for each. $\overline{c}\overline{c}$	No penalty for Mr. 2. Maximum penalty for Mr. 1. $\overline{c}c$
	confesses	Maximum penalty for Mr. 2. No penalty for Mr. 1. $c\overline{c}$	Fairly severe penalty for each. cc

Seeking to minimize the penalties they will suffer, the prisoners rank the four outcomes as follows:

Mr. 1	Mr. 2
$c\bar{c}$	$\bar{c}c$
$\overline{cc}$	$\overline{cc}$
cc	cc
$\bar{c}c$	$c\bar{c}$

Because $\bar{c}c$ and $\overline{cc}$ differ only in Mr. 2's action, he can block or overturn $\overline{cc}$ in favor of $\bar{c}c$, which he prefers to $\overline{cc}$. So $E_1(\{2\}, \bar{c}c, \overline{cc}, A)$ and thus $\bar{c}c \Delta \overline{cc}$. For similar reasons, $E_1(\{1\}, cc, \bar{c}c, A)$, $E_1(\{2\}, cc, c\bar{c}, A)$, and $E_1(\{1\}, c\bar{c}, \overline{cc}, A)$. Each prisoner is better off confessing, regardless of what the other does. Barring cooperation, then, each will confess: $\Gamma(\emptyset) = \{cc\}$. But both prisoners prefer $\overline{cc}$ to cc. So if cooperation is possible—if Messrs. 1 and 2 have the joint power to block cc in favor of $\overline{cc}$—then $E_1(\{1, 2\}, \overline{cc}, cc, A)$, and thus the four outcomes are related this way:

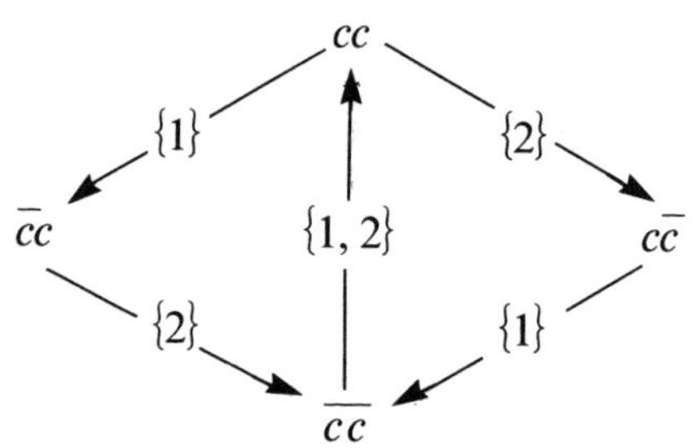

$\overline{cc}$, the one outcome for which cooperation is essential, is unstable—as the theorem says it must be. Specifically, either prisoner, acting alone, has the power to block or overturn $\overline{cc}$ in favor of another feasible outcome that he likes more.

Note that the E_1 structure of this example is the same as that of example C. And as in example C, the instability of cc does not demonstrate the impossibility of cooperation: an agreement not to confess might be enforceable.

Thing 6 Those of us who theorize about political economy have long been concerned with two phenomena, seemingly unconnected, both styled "collective irrationality." One is the instability of majority-voting and other collective-choice outcomes, as exemplified by the Classical Voting Paradox. The other is the failure of actors with partly conflicting preferences to cooperate in mutually advantageous ways, as in Prisoners' Dilemma games, free-rider problems, and Hobbesian states of nature. Theorem 11.2.1 subsumes the second variety of "collective irrationality" under the first, in this sense: If, to achieve a particular outcome, it is necessary to solve the cooperation problem, then doing so automatically gives rise to the stability problem. In other words, if cooperation by actors with partly conflicting preferences be essential to an outcome, that outcome must be unstable—which helps explain the difficulty of securing cooperation.

Thing 7 Generalized exchange need *not* involve *cooperation* in the strong sense of enforcing an agreement or joint course of action that is individually disadvantageous to those involved. The effect of trading votes across issues can be achieved, for example, by writing package (omnibus) legislation to begin with or by floor amendments.

Thing 8 Theorem 11.2.1 is completely trivial. What makes this noteworthy is that in institutional as opposed to strictly mathematical terms, Theorem 11.2.1 is considerably stronger than the impossibility theorems of sections 3.5 and 3.6. Under certain broadly applicable assumptions, Theorems 3.5.1, 3.5.2, and 3.6.1 show that the collective choice *can* be unstable—depending on the preference profile and feasible set, in the case of Theorems 3.5.1 and 3.5.2, and on the profile alone in the case of Theorem 3.6.1. Under different but also broadly applicable assumptions, Theorem 11.2.1 shows that the collective choice *must* be unstable—*regardless of the feasible set or preference profile.*

11.4. DECENTRALIZATION, COOPERATION, AND SOCO 2

Often collective choices are *decentralized* in the sense that different individuals or subgroups control different issues. A group *g* *controls* a set *d* of issues provided *g* has the power to move from any alternative to any other whenever the two alternatives differ only in the issues constituting *d*. *Cooperation* is possible when distinct groups who control distinct sets of issues also jointly control the combined set of issues. Both cooperation and its absence have interesting effects on collective choices.

The Concept of Cooperation

Control over sets of issues may be defined in terms of *POW*:

DEFINITION g controls d in α *iff d is a nonempty subset of* $\{1, 2, \ldots, m\}$, $\alpha \in S$, and *POW*(g, x, y, α) *for all distinct* $x, y \in \alpha$ *for which* $\{i \mid x$ and y have different ith coordinates$\} \subseteq d$.

This definition has two useful consequences:

THEOREM 11.4.1. (a) *If g controls d in α and if e is a nonempty subset of d, then g control e in α.* (b) *Assuming Axiom POW 4, if g controls d in α and if $g \subseteq h \subseteq N$ then h control d in α.*

Proof. Trivial. ■

For groups to cooperate, they must jointly control a combination of separately controlled issues. In the Prisoners' Dilemma the two prisoners will cooperate—neither will confess—only if they jointly control the set consisting of (1) the issue whether Mr. 1 confesses and (2) the issue whether Mr. 2 confesses. When a group g trades votes with a group h, agreeing to support $x_j \in A_j$ in return for h's support of $x_i \in A_i$, the separately controlled issues over which they thereby exercise joint control are not A_i and A_j; they are (1) the issue of how g will vote on A_j and (2) the issue of how h will vote on A_i.

The condition that cooperation is always possible may be formulated thus:

COOP 1 *If g controls d in α and h controls e in α then $g \cup h$ controls $d \cup e$ in α.*

Assuming Axiom POW 4, this is equivalent to

COOP 2 *If g controls d in α and g controls e in α, then g controls $d \cup e$ in α.*

THEOREM 11.4.2. *Assume Axiom POW 4. Then COOP 1 is equivalent to COOP 2.*

Proof. COOP 2 obviously is a special case (that in which $g = h$) of COOP 1. Now assume COOP 2. To deduce COOP 1, suppose g controls d in α and h controls e in α. Then by hypothesis and theorem 11.4.1(b), $g \cup h$ controls d in α and $g \cup h$ controls e in α. So by COOP 2, $g \cup h$ controls $d \cup e$ in α. ■

Whether or not cooperation is possible, SOCO 2 is an apparently reasonable choice-set characterization for decentralized collective-choice processes. I illustrate with four examples.

Economic Exchange

The first is example C of section 11.3—the banana-coconut exchange. The E_1 relationships within $\alpha = \{bc, b\bar{c}, \bar{b}c, \overline{bc}\}$ are the following:

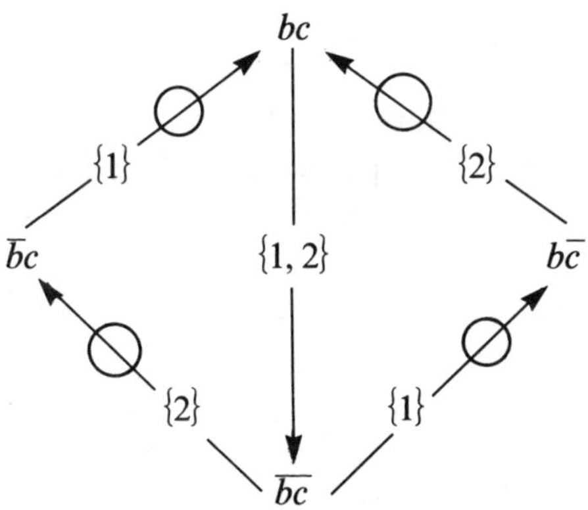

A circle marks each E_1 relationship for which *SUF* does not hold: Mr. 1 cannot secure $\bar{b}c$ because Mr. 2 is able and willing to move from $\bar{b}c$ to $\overline{bc}$, and $\overline{bc}$ is not the best Mr. 2 can do because $E(\{1, 2\}, bc, \overline{bc}, \alpha)$; similarly, Mr. 2 cannot secure $b\bar{c}$, and $\overline{bc}$ is not the best Mr. 1 can do. The E_2 relationships within α are given by all irreflexive chains of arrows that begin with the one uncircled arrow. So D_2^α has the form shown in figure 11.2. Therefore, SOCO 2 implies that $C(\alpha) = \{\overline{bc}\}$: Messrs. 1 and 2 will trade.

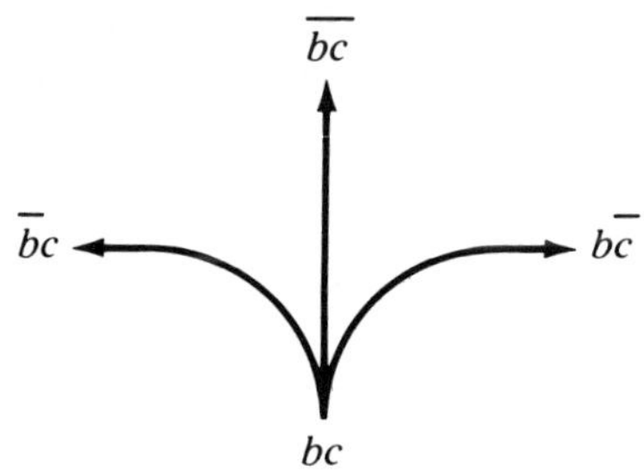

Figure 11.2

Prisoners' Dilemma

The second example is the Prisoners' Dilemma. Assuming cooperation is possible, the E_1 relationships within $\beta = \{cc, c\bar{c}, \bar{c}c, \overline{cc}\}$ are:

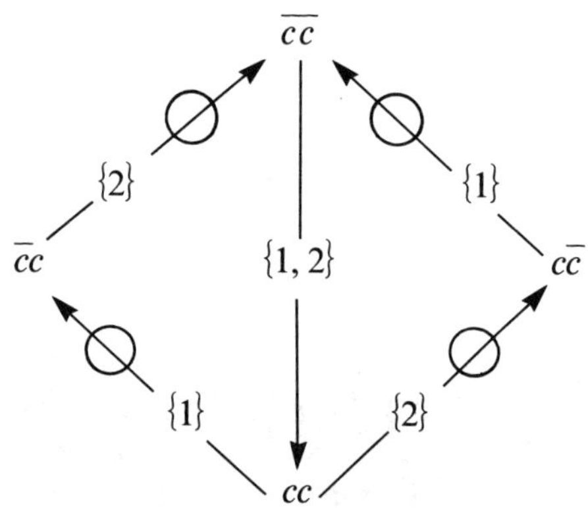

Structurally, this example is identical to the banana-coconut exchange. SOCO 2 implies that $C(\beta) = \{\overline{cc}\}$: the prisoners will not confess. Assume, on the other hand, that cooperation is *not* possible: not $POW(\{1, 2\}, \overline{cc}, cc, \beta)$. Then the E_1 relationships are the following:

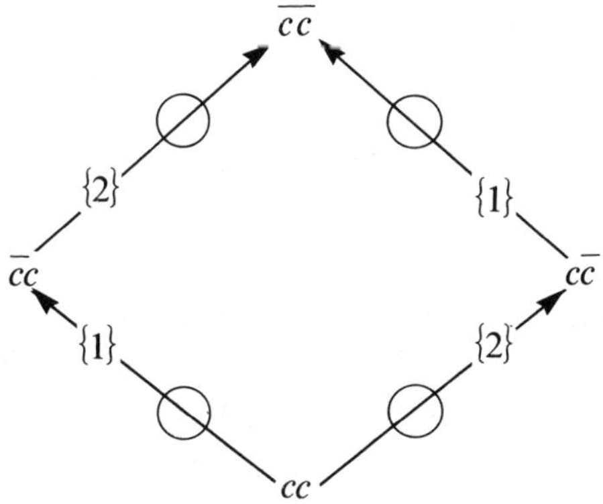

So D^β_2 has the form shown in figure 11.3. According to SOCO 2, then, $C(\beta) = \{cc\}$: both prisoners will confess.

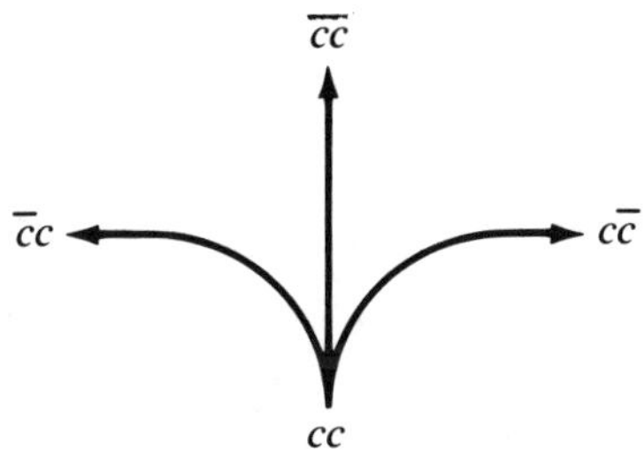

Figure 11.3

Sen's Liberal Paradox

The third example is Amartya Sen's celebrated Liberal Paradox. There are possible preference profiles under which Pareto Preference and Collective P-Acyclicity conflict with a condition Sen calls *Minimum Liberalism:* There are two individuals, Messrs. j and k, such that Mr. j has the absolute right to dictate the collective choice between two alternatives, say x and y, and Mr. k also has the absolute right to dictate the collective choice between two alternatives, z and w.

The conflict is obvious when $\{x, y\} = \{z, w\}$. So suppose $\{x, y\} \neq \{z, w\}$. If $\{x, y\}$ and $\{z, w\}$ overlap in one member—if, let us say, $x = z$—consider a profile of the form:

j	k	$N - \{j, k\}$
$x(z)$	y	y
y	w	w
w	$x(z)$	

By virtue of Minimum Liberalism, w is collectively preferred to x, and x to y. But by Pareto Preference, y is collectively preferred to w, contrary to Collective P-Acyclicity. Now suppose that $\{x, y\}$ and $\{z, w\}$ do not overlap at all. Consider a profile of the form:

j	k	$N - \{j, k\}$
w	y	
x	z	w and y
y	w	x z
z	x	

By Minimum Liberalism, x is collectively preferred to y, and z to w. But by Pareto Preference, w is collectively preferred to x, and y to z, contrary to Collective P-Acyclicity.

Only those who are religiously devoted to Collective P-Acyclicity, which causes trouble enough in other contexts, will find anything disturbing or paradoxical in Sen's result. Still it would be interesting to see what SOCO 2 says about situations of the type Sen envisages. It may help to begin by putting some flesh on the bare bones of Sen's two preference profiles.

For the first profile, suppose (to adapt an example given by Sen himself) that a single, publicly owned copy of *Lady Chatterly's Lover* is to be read by lascivious Mr. 1, by prudish Mr. 2, or by neither; call these feasible outcomes $l\bar{p}$, $\bar{l}p$, and $\overline{l\,p}$. Although Mr. 1 would enjoy reading the book, the thought of prudish Mr. 2 reading it and being shocked or titillated pleases him more. Mr. 2 feels that no one should read the book but that if anyone is going to read it then better he than corruptible Mr. 1. So Messrs. 1 and 2 rank the three outcomes as follows:

Mr. 1	Mr. 2
$\bar{l}p$	$\bar{l}\bar{p}$
$l\bar{p}$	$\bar{l}p$
$\bar{l}\bar{p}$	$l\bar{p}$

Either individual has the right to read or not to read the book as he chooses, so long as the other is not reading it—a modest "liberal" assumption. Where $\alpha = \{l\bar{p}, \bar{l}p, \bar{l}\bar{p}\}$, this means $POW(\{1\}, l\bar{p}, \bar{l}\bar{p}, \alpha)$ and $POW(\{2\}, \bar{l}\bar{p}, \bar{l}p, \alpha)$. Consequently, $E_1(\{1\}, l\bar{p}, \bar{l}\bar{p}, \alpha)$ and $E_1(\{2\}, \bar{l}\bar{p}, \bar{l}p, \alpha)$. And if cooperation is possible—hence if $POW(\{1, 2\}, \bar{l}p, l\bar{p}, \alpha)$—then $E_1(\{1, 2\}, \bar{l}p, l\bar{p}, \alpha)$. So the E_1 relationships within α are either

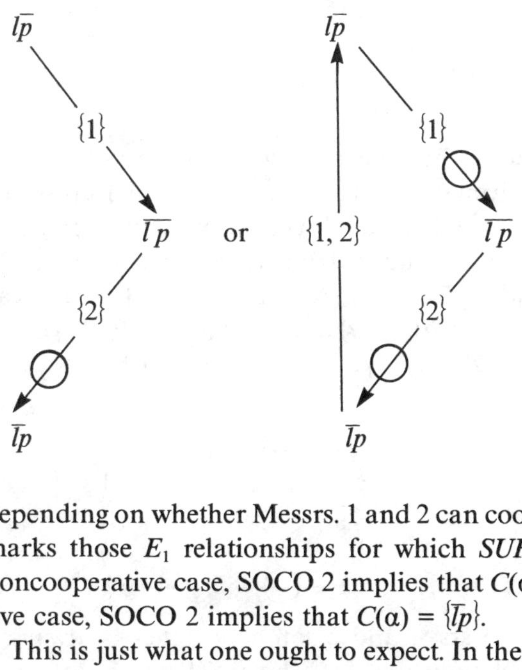

depending on whether Messrs. 1 and 2 can cooperate. As usual, a circle marks those E_1 relationships for which *SUF* does not hold. In the noncooperative case, SOCO 2 implies that $C(\alpha) = \{l\bar{p}\}$. In the cooperative case, SOCO 2 implies that $C(\alpha) = \{\bar{l}p\}$.

This is just what one ought to expect. In the absence of cooperation, were Mr. 2 given the book, he would keep it without reading it. But that would violate Mr. 1's right to read it if no one else is reading it. So Mr. 1 ought to be given the book. Because he cannot cooperate with Mr. 2, Mr. 1 will keep the book and read it. But Messrs. 1 and 2 both prefer Mr. 2's reading the book ($\bar{l}p$) to Mr. 1's reading it ($l\bar{p}$). So if cooperation is possible, Messrs. 1 and 2 will secure $\overline{l}p$: they will reach an enforceable agreement to the effect that Mr. 2 should read the book.

For Sen's second profile, suppose the city government is to place an ugly utility pole in Mr. 1's backyard ($b1$), in Mr. 1's front yard ($f1$), in Mr. 2's backyard ($b2$), or in Mr. 2's front yard ($f2$). Each of these worthies occasionally sits in his backyard, spends virtually no time in front of his house, and frequently sits at his front window, from which he can see the other's front yard but not his own. Because each want to minimize the time he spends seeing the pole, Messrs. 1 and 2 rank the four outcomes thus:

Mr. 1	Mr. 2
$b2$	$b1$
$f1$	$f2$
$b1$	$b2$
$f2$	$f1$

Although neither individual has the right to keep the pole off his property, each has been given the right to decide where on his property to place the pole if it is placed there at all. That is, where $\gamma = \{f1, f2, b1, b2\}$, we have $POW(\{1\}, f1, b1, \gamma)$ and $POW(\{2\}, f2, b2, \gamma)$. Consequently, $E_1(\{1\}, f1, b1, \gamma)$ and $E_1(\{2\}, f2, b2, \gamma)$. And if Messrs. 1 and 2 can cooperate, hence if $POW(\{1, 2\}, b1, f2, \gamma)$ and $POW(\{1, 2\}, b2, f1, \gamma)$, then $E_1(\{1, 2\}, b1, f2, \gamma)$ and $E_1(\{1, 2\}, b2, f1, \gamma)$. So the E_1 relationships within γ are either

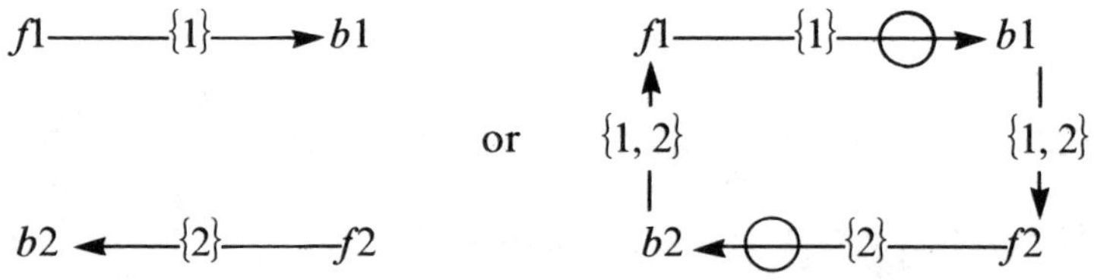

depending on whether Messrs. 1 and 2 can cooperate. As always, a circle marks those E_1 relationships for which SUF does not hold. Therefore, SOCO 2 implies that $C(\gamma) = \{b1, b2\}$ in the cooperative case and $C(\gamma) = \{f1, f2\}$ in the noncooperative case.

In the book and pole examples, the cooperative outcome was unstable in the E_1 sense. That was inevitable, and it had nothing to do with liberalism or individual rights: to cooperate is to engage in generalized exchange, and outcomes for which generalized exchange is essential are perforce unstable, whether or not they are based on individual rights or liberal principles. In short, Sen's Liberal Paradox is just another instance of Theorem 11.2.1. Although instability is virtually inescapable, however, that presents no theoretical problem: SOCO 2 specifies a reasonable solution to collective-choice problems in the face of instability.

Cooperation Problems Under Majority Rule

The final example involves voting and a majoritarian power structure. Suppose three voters assign dollar values to three bills, *a, b,* and *c*, as follows:

	Mr. 1	Mr. 2	Mr. 3
a	4,000,000	4,000,000	−9,000,000
b	−9,000,000	4,000,000	4,000,000
c	4,000,000	−9,000,000	4,000,000

In effect, each bill prescribes a transfer of priceable assets from one voter to the others, and perhaps a transformation of those assets from money into other forms. In each case, the cost to the nonbeneficiary, measured in dollars, is greater than the total benefit to the beneficiaries, owing to transfer costs. Each bill is voted up or down by Simple Majority Rule.

The passage of each bill is preferred to its defeat by some majority. So in the absence of cooperation, each bill would pass, as a result of which everyone would loose $1 million.

If Messrs. 1, 2, and 3 cooperate by trading votes, each agreeing to vote against the two bills he favors in return for a like agreement by the other two voters, all three bills will be defeated and no one will lose money.

If Messrs. 1 and 2 alone trade votes, they can do even better for themselves: they can ensure passage of *a* and defeat of *b* and *c*. The result would be a gain of $4 million for each of them—and a loss of $9 million for Mr. 3.

It is worth noting that each of the bills might itself represent a cooperative outcome. For example, *a* might be the result of a vote-trade agreement between Messrs. 1 and 2 to support $a_1 = (8{,}500{,}000, -4{,}500{,}000, -4{,}500{,}000)$ and $a_2 = (-4{,}500{,}000, 8{,}500{,}000, -4{,}500{,}000)$, *b* the result of a similar agreement between Messrs. 2 and 3, and *c* the result of a similar agreement between Messrs. 1 and 3. This shows that successful cooperation can engender new cooperation problems at a higher level.

Such problems can be avoided by forming an antifactional universal coalition—representing the "general will" or "public interest"—or by forming a disciplined majority party: in the example, the universal coalition of Messrs. 1, 2, and 3 can ensure that no one looses anything, and the majority "party" of Messrs. 1 and 2 can secure great rewards for its members at considerable cost to the nonmember. To do this, Messrs. 1 and 2 must stick together on all issues—unlike the majorities who coalesced just to support the *a* package, the *b* package, and the *c* package. Such is the essential difference between a disciplined political party and a vote-trade coalition.

The eight possible outcomes constitute a cycle. It is hard to say which will be chosen. That depends on which majority succeeds in coalescing. But it is clear that if cooperation is possible then *abc* will *not* be chosen. This is just what SOCO 2 says: the choice set comprises all eight outcomes other than *abc*.

11.5. COLLECTIVE CONTROL AND INDIVIDUAL CHOICE

In my discussion of decentralized collective-choice processes, I interpreted the issues $A_1, \ldots, A_m$ as feasible sets open to individuals or to subgroups of *N*. When examining *centralized* collective-choice processes, too, it is often illuminating to interpret $A_1, \ldots, A_m$ as feasible sets open to individuals—the same individual at different times or different individuals at the same or different times. For one of the most conspicuous objects of collective choice is the regulation of individual choice—the collective adoption of policies, practices, and rules with which individuals are required to comply. Difficulties arise, though, when we try to say what policies, practices, and rules are and what it is for individuals to comply with them.

Suppose we want individuals to comply with some simple social policy, say the classical *utilitarian* injunction to maximize aggregate happiness. The problem is that individual choices which singly maximize aggregate happiness might not jointly maximize aggregate happiness, witness these two examples:

Example E Dr. Krankheit neglected to perform a much-needed surgical operation for which he had been hired. As a result, the patient died and Dr. Krankheit was hauled before a medical-society review board, where he defended himself as follows:

"I first had to choose between administering and not administering an anesthetic—between a and $\bar{a}$, let us say. I chose $\bar{a}$. I then had to choose between performing the surgery itself and not performing it—between s and $\bar{s}$. Such were my firm intentions—you can check this with a lie detector, if you like—that I would not have chosen s even had I chosen a. So a would have done no good. Therefore, because a was costly in several obvious ways, my choice if $\bar{a}$ from $\{a, \bar{a}\}$ maximized aggregate happiness."

"But why," he was asked by a board member, "did you not perform the *surgery*?"

"Good heavens!" exclaimed Dr. Krankheit, with just the right mix of incredulity and indignation. "Operate on a nonanesthetized patient? That would have been inhumanly cruel."

Example F A child-eating monster has attacked a village in a valley, captured all the children, and confined them at the foot of a cliff in two pens, between which he currently sits, decanting a jeroboam of Château La Tour, 1959, while waiting for a large kettle to come to the boil. At the top of the cliff stand Messrs. 1 and 2, along with a large boulder. If both push the boulder, it will fall off the cliff and mash the monster, saving all the children. If neither pushes, the boulder will stay where it is and the monster will eat half the children. If either pushes by himself, the boulder will go off the cliff, but at the wrong angle to mash the monster; it will land instead in one of the pens, killing half the children, leaving the remaining half for the monster to eat.

Messrs. 1 and 2 each has to choose between pushing and not pushing the boulder—between p_1 and $\bar{p}_1$ for Mr. 1, between p_2 and $\bar{p}_2$ for Mr. 2. As it happens, neither of them pushes: the actual outcome is $\bar{p}_1\bar{p}_2$. And such are their firm intentions that neither of them *would* have pushed, even had the other pushed. Consequently, p_1 would have resulted in the death of *all* the children, compared with half for $\bar{p}_1$, so $\bar{p}_1$

maximized aggregate happiness. Likewise for $\bar{p}_2$. From a classical utilitarian point of view, Messrs. 1 and 2 each did the right thing by not pushing.

In each example there were two issues. Although the choice from each issue maximized aggregate happiness, the combination of choices ($\overline{as}$ or $\bar{p}_1\bar{p}_2$) produced less aggregate happiness than an alternative combination (as or p_1p_2) would have produced.

Part of the reason for this is that the utilitarian preferability relation is *nonseparable* (or *complementary*): whether a position on an issue would produce more aggregate happiness than another position on the same issue sometimes depends on the positions chosen from other issues. It is quite common for individuals to have nonseparable preferences. Someone who prefers red wine to white with meat but white to red with fish has a preference that is not separable over wine and food issues: whether he prefers one wine to another depends on what he is eating.

So-called *rule utilitarians* would apply the utilitarian test, not directly to individual choices, but to rules or practices—represented by elements or subsets of A, perhaps. They contend that individuals should observe rules or practices whose universal observance would maximize aggregate happiness.

But that runs afoul of this variant of example F: As before, such are Mr. 1's firm intentions that he would not push even if Mr. 2 pushed. But in this example Mr. 2 is differently disposed: *he would* push *if Mr. 1 pushed.* Because universal observance of the practice of pushing in such cases would maximize aggregate happiness, Mr. 2 ought to push according to the rule-utilitarian principle. But because Mr. 1 would not push even if Mr. 2 pushed, Mr. 2's observance of the rule-utilitarian prescription would result in the death of all the children rather than half of them.

In the examples, each individual maximized aggregate happiness in the actual cricumstances of his choice, but he was unwilling to do so in other possible circumstances. Because $\bar{a}$ was chosen, $\bar{s}$ maximized aggregate happiness. Had a been chosen, Dr. Krankheit would still have chosen $\bar{s}$, even though $\bar{s}$ would not have maximized aggregate happiness in that case. Because, in the original monster example, Mr. 2 chose $\bar{p}_2$ and would have done so regardless of Mr. 1's choice, Mr. 1's choice of $\bar{p}_1$ maximized aggregate happiness. Had Mr. 2 been disposed to choose p_2 on condition that Mr. 1 choose p_1, Mr. 1 still would have chosen $\bar{p}_1$, even though $\bar{p}_1$ would not have maximized aggregate happiness in that case.

I draw this lesson: An individual should be regarded as complying with the utilitarian policy only if his hypothetical choices as well as his actual ones maximize aggregate happiness. More generally, individuals comply with a policy or practice or rule, utilitarian or otherwise, only if their hypothetical as well as their actual choices conform to it. When a policy or practice or rule is adopted, the mechanism of enforcement—evaluation and instruction, praise and blame, punishment and reward—must apply to *all possible* choices by individuals, not just to the few choices they have the chance to make. (Of course, this makes the problem of enforcement more difficult.)

To make the suggestion precise, we must represent a policy or practice or rule by a *choice function*, say F, not by an element or subset of A. So $F: S \rightarrow S$ and $F(\alpha) \subseteq \alpha$ for all $\alpha \in S$. Messrs. 1, 2, . . . , n have their own choice functions, $F_1, \ldots, F_n$, on S. Mr. i *complies with* F iff, for every $\alpha \in S$ whose members differ only in dimensions controlled by $\{i\}$, $F_i(\alpha) \subseteq F(\alpha)$.

The failure of individual hypothetical choices to conform to a rule or policy is one reason individual choices—actual ones—can singly conform to the rule or policy while jointly violating it. There is another reason, illustrated by this example:

Example G Messrs. 1 and 2 share an island that has cars and roads but no rule of the road—no rule that prescribes driving only on the left or only on the right. If one of them drives on the left and the other on the right (*lr* or *rl*), there is a good chance they will collide. Both would be better off if both drove on the same side (*rr* or *ll*); it does not matter which side. So if F represents the utilitarian policy—or any of a number of other social policies—then $F(\{ll, lr, rl, rr\}) = \{rr, ll\}$. The trouble is that Mr. 1 might try to comply with F by driving on the left while Mr. 2 tries to comply with F by driving on the right. The result would be *lr*, which does not belong to the choice set.

The problem, more generally, is that Messrs. 1 and 2 must *coordinate to comply*. Formally speaking, the choice set $\{rr, ll\}$ is unsaturated, a *saturated* set being a set α with this virtue:

For all $x_1, \ldots, x_m$, if x_i is the ith coordinate of some vector in α, $i = 1, 2, \ldots, m$, then $(x_1, \ldots, x_m) \in \alpha$.

$\{rr, ll\}$ is not saturated because l is the first coordinate of ll, and r the second coordinate of rr, but $lr \notin \{ll, rr\}$.

Often choice sets have to be pared down, however arbitrarily, to ensure saturation. Thus, $\{ll, rr\}$ must be pared to $\{ll\}$ or $\{rr\}$. Rules that pare down initially unsaturated choice sets to saturated subsets are called *conventions*. The rule against driving on the left, for example, is a convention that pares $\{ll, rr\}$ to $\{rr\}$.

Unsaturated choice sets cause special problems because conventions are costly: they must be created by human beings, the need for them is sometimes hard to foresee, and to be effective they must be publicized and generally acknowledged.

SOURCES AND RELATED CONTRIBUTIONS

The general treatment of multistage choice processes has not previously been published. Campbell (1977) and Schwartz (1977) discuss the special case of *serial*-choice processes. Serial choice is related to the celebrated Path Independence condition of Plott (1973):

$$C(C(\alpha) \cup \beta) \equiv C(\alpha \cup \beta).$$

The intuitive idea is that the choice set ($C(\alpha \cup \beta)$) from a given set ($\alpha \cup \beta$) should not depend on the particular serial-choice path (choose from α, then from $C(\alpha) \cup \beta$) through which it was reached. See also Blau (1975), Parks (1976a), and Ferejohn and Grether (1977). My treatment of multistage choice processes also is related to the idea, which has received considerable attention recently, that political outcomes often are best explained in terms of the fine details of institutional processes and agenda structures; see especially Shepsle (1979) and Shepsle and Weingast (1981b).

WARP and Theorem 10.2.5 come from Arrow (1959). See also the related contributions of Chernoff (1954), Tullock (1964), Hansson (1968), Wilson (1970), Herzberger (1973), and Fishburn (1974b). Samuelson (1938) first formulated the Weak Axiom of Revealed Preference as a condition on demand functions defined for economic opportunity sets rather than on choice functions defined for all finite subsets of some arbitrary set of objects; see also Houthakker (1950, 1965). The set of equivalence theorems presented in section 10.2 comes from Schwartz (1976). For earlier, related treatments, see Sen (1970: chs. 1 and 1*; 1971) and Jamison and Lau (1973, 1975).

The results in section 10.4 about choice by specification were reported after 1975 in my *Notes on the Abstract Theory of Collective Choice*, various versions of which were published from 1970 onward at Stanford and Carnegie-Mellon.

Theorem 10.2.1, the Universal Instability Theorem, was presented in Schwartz (1981). This generalizes a series of previous results couched mainly in terms of voting and vote trading rather than collective choice and generalized exchange, which come from Downs (1957:55–69), Kadane (1972), Oppenheimer (1972, 1975), Bernholz (1973, 1974b, 1975), Koehler (1975), Miller (1975, 1977b), Schwartz (1976), and Enelow and Koehler (1975). It was the brilliant pioneer work of Bernholz and Oppenheimer that prompted my own thoughts on the subject. These results, like the "spatial" instability results reported under "Sources . . . " at the end of Part II, make severe assumptions about individuals' preferences, the collective-choice process, and especially the feasible set (which is not allowed to be an arbitrary subset of the cross-product of issues), precluding any inference about the prevalence of instability—and motivating Theorem 10.2.1, which does not rest on such assumptions.

Cooperation problems are discussed at a very general level and from a variety of points of view by von Neumann and Morgenstern (1953), Samuelson (1954), Margolis (1955), Luce and Raiffa (1957), Olson (1965), Howard (1971), Schwartz (1973: ch. 9), Taylor (1976), Frohlich and Oppenhiemer (1978), and Axelrod (1984).

The Liberal Paradox comes from Sen (1970b; 1970a; chs. 6 and 6*; 1976). For other interesting approaches, see Gibbard (1974), Bernholz (1974a, 1983), Fine (1975), Blau (1975), Osborn (1975), Kelley (1976a,b; 1978: ch. 9), and Aldrich (1977).

The compliance problem (the converse problem of collective choice) is presented, at greater length but less formally, in Schwartz (1985). The rock-and-monster example, however, comes from Schwartz (1973:145–156), where I used it to prove the nonequivalence act- and rule-utilitarianism. Allan Gibbard has independently concocted a similar example (down to some inessential dramatic details) to the same end (personal communication).

CONCLUSION

A picture of collective-choice processes, and more specifically of political institutions, emerges from the findings of these chapters. Based on an abstract, deductive inquiry, this picture is painted from too high an altitude to capture small details of the political landscape. But the colors and contours that it reveals are not likely to be so clearly depicted from a closer perspective.

Consistency conditions on choice functions require choices from different potential feasible sets to be consistent with one another in specified ways—to conform to specified patterns or regularities. They rule out certain combinations of choices from different sets. Because preference is pairwise choice, the "rationality" conditions of chapter 2—BICH, P-Acyclicity, and the varieties of Transitivity—are consistency conditions. So, of course, are the GETCHA and GOCHA conditions of chapter 6. And so, most obviously, are the insensitivity conditions of chapter 10—WARP and variations. Although they simplify the theory of collective choice in intuitively appealing ways, these conditions are all pregnant with paradox.

Because the classical "rationality" conditions, even BICH and P-Acyclicity, conflict with modest requirements of other sorts, collective-choice processes cannot be assumed to produce best or maximal or stable choices. As you saw in Part II, these conflicts are not peculiar to democratic choice processes, nor to choice processes fulfilling stringent informational constraints (independence of preferences regarding infeasible alternatives and of interpersonal comparisons of preference intensity). They are not even peculiar to collective, as distinct from individual, choice processes. And as you saw in chapter 11, instability is no mere theoretical possibility: it arises whenever collective choices depend on generalized exchange, regardless of preference profile, choice process, or feasible set.

Because the "rationality" conditions, from weakest to strongest, are equivalent to insensitivity conditions (section 10.2), collective choices are highly sensitive to which alternatives are feasible. What makes this especially interesting is that the feasible set is a kind of decision variable—something chosen by us rather than a datum given by nature (section 10.1).

Although GETCHA and GOCHA are in some ways less demanding than the classical conditions of "rationality," they specify choice sets that can be large and Pareto inefficient. Because these sets comprise just the alternatives chosen by innumerable multi-stage processes and by the process of choice by specification, depending chiefly on agenda structure, such processes are quite indeterministic—or, rather, their outcomes depend more heavily than one might have supposed on certain apparently "arbitrary" factors.

None of this means that collective choices are hopelessly inexplicable. One thing it does mean is that the "rationality" and other consistency conditions purchased a kind of mathematical simplicity at too high a price. But SOCO 2, which unified much of our picture of collective choice, is not a consistency condition. Our findings also mean that collective choices are far more sensitive than some theorists may have supposed to agenda features—to the feasible set and its internal organization—which are themselves a matter of choice. But that just shows we need to know a good deal of contextual detail in order to explain a given choice. Our picture is not one of ineffable complexity. It is just less simple and deterministic than the picture that led to paradoxes. Like the real world, it is not so much *inexplicable* as it is *interesting*.

I want to conclude with a closer look at one part of this picture, which some may judge to be unfaithful to the real world, hence to be a paradox as yet unresolved. Collective-choice processes cannot be regarded as producing maxima or stable outcomes of any sort—outcomes undominated under some appropriate binary relation. But that is no problem. Those disturbed by the instability of collective choices are disturbed, I believe, because they would like to predict and explain political outcomes and because they do not perceive the political world to be utterly chaotic. But instability does not preclude prediction: SOCO 2 spells out a reasonable solution concept applicable in the face on instability—one that might be further refined for special contexts. Nor do instability results show the world to be chaotic. The instability revealed by Theorem 10.2.1 is an instability *in prospect* but not necessarily *in retrospect.* Once a collective choice has been made, an enforcement mechanism (legal coercion, conscience, a norm of reciprocity, or whatnot) often takes over, changing the original power structure so that those who had the power and incentive to overturn the chosen outcome no longer have it. In the first of two periods, no outcome is stable under the prevailing power structure, but a choice is made according to SOCO 2. In the second period, the power structure is changed so that the outcome chosen in the first period is now stable. Thanks to SOCO 2, prospective instability poses no problem for prediction. And retrospective instability is not inevitable and might not be a frequent occurrence.

No doubt retrospective instability is sometimes absent and sometimes fragile and fleeting. But what of it? The political world we observe is not frozen with stability any more than it is fluid with chaos. Nor is it the least bit obvious that stability in politics is, without qualification, a Good Thing. Why should political processes move to points of no change? Let me suggest that it is a virtue of political systems to institutionalize such shifting of alignments or governmental change as otherwise would occur in a more violent, less predictable, and, therefore, more costly manner. Like a well-designed, well-managed railroad, a well-ordered polity limits and channels movement and provides numerous, clearly marked points of departure and destination but does not permanently cease to run once some particular station has been reached.

APPENDIX

TERMS AND SYMBOLS OF SET THEORY

The mathematical terms and symbols used in this book are mostly set theoretic. This survey of terms and symbols is intended as a review and reference list for readers who have had some experience with elementary intuitive set theory. Others should consult a textbook on the subject, such as Suppes (1960). Note that synonyms are grouped together in the second column.

Category	Set-Theoretic Expression	Explanation
Relations between sets	$x \in y$ x is a *member* of y x is an *element* of y x *belongs* to y	y is a set of objects, one of which is x. *Example:* $3 \in$ set of integers. Sometimes "$\in$" is used as a preposition, as in "Every $x \in \alpha$ is divisible by 4," meaning "Every x *in* α (*belonging to* α) is divisible by 4."
	$x_1, \ldots, x_k \in y$	$x_1 \in y$ and $x_2 \in y$ and ... and $x_k \in y$. *Example:* 1, 13, 84 $\in$ set of integers.
	$x \subseteq y$ x is a *subset* of y y is a *superset* of x	Every member of x is a member of y. *Example:* Let I be the set of all integers and E the set of even inte-

Category	Set-Theoretic Expression	Explanation
		gers; then $E \subseteq I$. Note the difference between subsets and members: $3 \in I$ but $3 \nsubseteq I$, and $E \subseteq I$ but $E \notin I$. Note, too, that every set is a subset of itself.
	$x \subset y$ x is a *proper* subset of y	$x \subseteq y$ but $y \nsubseteq x$. Example: $E \subset I$.
	x and y are *disjoint* sets	x and y are sets with no members in common.
	x is a *partition* of y	x is a set (or *family*) of mutually disjoint subsets of y, and every member of y belongs to some member of x.
Descriptions of sets in terms of their members	$\{x_1, \ldots, x_k\}$	The set whose members are $x_1, \ldots, x_k$ and nothing else. *Example:* {1, 847, Kenneth Arrow} = the set consisting of the numbers 1 and 847 and Kenneth Arrow.
	$\{x, y\}$ the *pair set* of x and y the *unordered pair* of x and y	The set whose sole members are x and y. It is "unordered" because $\{x, y\} = \{y, x\}$.
	$\{x\}$ the *unit set* of x	The set whose sole member is x.
	$\{x \mid \ldots x \ldots\}$	The set of objects x such that $\ldots x \ldots$. *Example:* $\{x \mid x$ is a dog$\}$ = the set of dogs.

Category	Set-Theoretic Expression	Explanation
	$\{x \in \alpha \mid \ldots x \ldots\}$	The set of those objects *x in* α such that . . . *x* . . . . *Example:* Let *I* be the set of integers; then $\{x \in I \mid 3 \leq x \leq 817\}$ = the set of integers between 3 and 817.
	$\emptyset$ the *empty* set the *null* set	The (one and only) set with no members. Note that $\emptyset$ is a subset of every set.
Operations on sets	$x \cap y$ the *intersection* of *x* and *y*	$\{z \mid z \in x \textit{ and } z \in y\}$
	$x \cup y$ the *union* of *x* and *y*	$\{z \mid z \in x \textit{ or } z \in y\}$
	$x - y$ the (relative) *complement* of *y* with respect to *x*	$\{z \mid z \in x \text{ but } z \notin y\}$
	$\bigcup \alpha$ the *union over* α	The union of all the sets belonging to α. $\{x \mid x \in \beta$ for some $\beta \in a\}$. *Example:* $\bigcup \{\{2, 3\}, \{1, 8, 4\}, \{5, 6, 7, 9, 2\}\} = \{1, 2, 3, 4, 5, 6, 7, 8, 9\}$.
	$\bigcup_{i=1}^{k} \alpha_i$	$\alpha_1 \cup \alpha_2 \cup \cdots \cup \alpha_k$
	$\mathscr{P}(\alpha)$ 2^α *the power set* of α	The set of all subsets of α. $\{\beta \mid \beta \subseteq a\}$. *Example:* $\mathscr{P}(\{1, 13\}) = \{\emptyset, \{1\}, \{13\}, \{1,13\}\}$.
Vectors	(x, y) the *ordered pair* of *x* and *y*	The combination of *x* and *y in that order.* In $\{x, y\}$, order does not count: $\{x,$

Category	Set-Theoretic Expression	Explanation
		$y\} = \{y, x\}$. In (x, y), order counts: $(x, y) \neq (y, x)$ (unless $x = y$).
	$(x_1, \ldots, x_k)$ the *ordered k-tuple* of $x_1, \ldots, x_k$ the k-dimensional *vector* of $x_1, \ldots, x_k$	The combination of $x_1, \ldots, x_k$ in that order.
	the ith *coordinate* of vector v	The ith component (from left to right) of v.
	$\{(x, y) \mid \ldots x \ldots y \ldots\}$	The set of ordered pairs (x, y) such that $\ldots x \ldots y \ldots$. *Example:* $\{(x, y) \mid x$ and y are integers and $y = x + 1\}$ = the set of ordered pairs of an integer followed by its successor.
	$\alpha \times \beta$ *Cartesian product* of α and β *cross product* of α and β	$\{(x, y) \mid x \in \alpha$ and $y \in \beta\}$
	$\bigtimes_{i=1}^{k} \alpha_k$	$\alpha_1 \times \alpha_2 \times \cdots \times \alpha_k$
	α^k	$\underbrace{\alpha \times \alpha \times \cdots \times \alpha}_{k \text{ times}}$

Category	Set-Theoretic Expression	Explanation
Relations	*binary relation*	Set of ordered pairs. *Examples:* $\{(x, y) \mid x$ is the mother of $y\}$ = motherhood, $\{(x, y) \mid x$ and y are integers and $x > y\}$ = relation of greater to less among integers.
	xRy *x bears R* to *y*	$(x, y) \in R$. When R is a binary relation, we usually write xRy instead of $(x, y) \in R$. *Example:* R = motherhood, so xRy iff $(x, y) \in R$ iff x is y's mother.
	k-ary *relation*	Set of ordered k-tuples. *Example:* $\{(x, y, z, w) \mid x$ is the father and y the mother of z, who is a boy, and w, a girl$\}$ = the quaternary relation of a father and mother to their son and daughter.
	$R(x_1, \ldots, x_k)$	$(x_1, \ldots, x_k) \in R$. When R is a k-ary relation, for $k > 2$, we usually write $R(x_1, \ldots, x_k)$ instead of $(x_1, \ldots, x_k) \in R$.
	R is a k-ary *relation on* α	$R \subseteq \alpha^k$.
	the *domain* of a binary relation R	$\{x \mid xRy$ for some $y\}$
	the *range* of a binary relation R	$\{x \mid yRx$ for some $y\}$
	the *converse* of a binary relation R	$\{(x, y) \mid yRx\}$

Category	Set-Theoretic Expression	Explanation
	the *inverse* of R R^{-1}	
Order-relevant properties of a binary relation R	R is *reflexive* in α	xRx for all $x \in \alpha$.
	R is *irreflexive* in α	xRx for no $x \in \alpha$.
	R is *symmetric* in α	For all $x, y \in \alpha$, if xRy then yRx.
	R is *asymmetric* in α	For all $x, y \in \alpha$, if xRy then not yRx.
	R is *antisymmetric* in α	For all $x, y \in \alpha$, if xRy and yRx then $x = y$.
	R is *connected* in α	For all $x, y \in \alpha$, if $x \neq y$ then either xRy or yRx.
	R is *strongly connected* (complete) in α	For all $x, y \in \alpha$, either xRy or yRx.
	R is *transitive* in α	For all $x, y, z \in \alpha$, if xRy and yRz then xRz.
	R is *negatively transitive* in α	For all $x, y, z \in \alpha$, if not xRy and not yRz then not xRz.
	R is *intransitive* in α	For all $x, y, z \in \alpha$, if xRy and yRz then not xRz.
	R is *nontransitive* in α	There exist $x, y, z \in \alpha$ such that xRy and yRz but not xRz.
	R is *acyclic* in α	There exist no k and $x_1, \ldots, x_k \in \alpha$ such that $x_1Rx_2R \cdots Rx_kRx_1$.

Category	Set-Theoretic Expression	Explanation
	R is a *suborder* of α	R is asymmetric and acyclic in α.
	R is a *quasiordering* (preordering) of α	R is reflexive and transitive in α.
	R is a (reflexive) *partial ordering* of α.	R is reflexive, antisymmetric, and transitive in α.
	R is a *strict partial ordering* of α	R is asymmetric and transitive in α.
	R is a (reflexive) *weak ordering* (complete ordering) of α	R is transitive and strongly connected in α.
	R is a *strict weak ordering* (a weak ordering in the asymmetric sense, used here) of α	R is asymmetric and negatively transitive in α.
	R is a *linear* (simple) *ordering* (in the reflexive sense) of α	R is antisymmetric, transitive, and strongly connected in α.
	R is a *linear ordering* (in the strict sense, used here, a.k.a. a strict simple ordering) of α	R is asymmetric, transitive, and connected in α.
Functions	A *unary function* (operation, transformation, mapping)	A binary relation f such that, for all x, y, z, when xfy and xfz then $y = z$. The condition means that no one thing bears the relation f to two things.

Category	Set-Theoretic Expression	Explanation
	$f(x) = y$ y is the *value* of f at x	$(x, y) \in f$ (assuming f is a unary function). Because there is just one object to which x bears the relation f, we can refer to this object, unambiguously, as *the* f- relatum of x, or simply $f(x)$.
	k-ary *function*	$(k + 1)$-ary relation f such that, for all $x_1, \ldots, x_k, y, z$, if $f(x_1, \ldots, x_k, y)$ and $f(x_1, \ldots, x_k, z)$ then $y = z$.
	$f(x_1, \ldots, x_k) = y$ y is the *value* of f *at* $(x_1, \ldots, x_k)$	$(x_1, \ldots, x_k, y) \in f$ (assuming f is a k-ary function).
	the *domain* of a k-ary function f	$\{(x_1, \ldots, x_k) \mid f(x_1, \ldots, x_k) = y$ for some $y\}$
	f is *on* α f is *defined on* α	α is the domain of f.
	f is *defined for* $(x_1, \ldots, x_k)$	$(x_1, \ldots, x_k)$ belongs to the domain of f.
	the *range* of a k-ary function f	The set of f's values. $\{y \mid y = f(x_1, \ldots, x_k)$ for some $x_1, \ldots, x_k\}$
	f is *into* α	The range of f is a subset of α.
	f is *onto* α	The range of f *is* α.
	$f: \alpha \to \beta$ f maps α into β	f is a function on α and into β.

Category	Set-Theoretic Expression	Explanation
	the *image* of the set α under the function *f*	$\{x \mid f(y) = x \text{ for some } y \in \alpha\}$
	a *one-to-one* function	A function *f* such that, for all *x*, *y*, *z*, if $f(x) = y$ and $f(z) = y$ then $x = z$. A *function* relates each thing to at most one thing. A *one-to-one* function has the additional feature that it relates at most one thing to each thing. If *f* is one-to-one, f^{-1} (the inverse of *f*) also is a function.
	a *permutation* of α	A one-to-one function on α and onto α.
Numerical concepts	$\mid\alpha\mid$ the *cardinal number* of α the *cardinality* of α	The number of members of α.
	α and β have the *same cardinality*	There exists a one-to-one function on α onto β.
	$\sum_{i=1}^{k} x_i$	$x_1 + x_2 + \cdots x_k$
	$\sum_{x \in \alpha} f(x)$	The sum of numbers $f(x)$ for all $x \in \alpha$. When reckoning this sum, we count a number *k* as many times as there are members *x* of α for which $k = f(x)$.
	max(k, m)	The larger of *k* and *m*.
	inductive proof	To prove by *mathematical*

Category	Set-Theoretic Expression	Explanation
	proof *by induction*	*induction* that every natural number (nonnegative integer) has property ϕ, we first prove $\phi(0)$, then assume $\phi(k)$ (this assumption is our "inductive hypothesis") and deduce $\phi(k + 1)$. To prove by *complete* or *course-of-values* induction that $\phi(k)$ for every natural number k, we assume $\phi(j)$ for *all* $j < k$ (our inductive hypothesis) and deduce $\phi(k)$. Sometimes we give an inductive proof of a theorem ostensibly about objects other than numbers. To prove, for example, that every finite set α of a certain sort has property ψ, we might argue "by induction on $\|\alpha\|$," assuming that every smaller set has ψ and deducing $\psi(\alpha)$. This is really a special case of the above pattern because, in proving that every set of the given sort has property ψ, we thereby prove inductively that every natural number k has the following, more complex property: all sets of the given sort with cardinality k have property ψ.

REFERENCES AND SELECTED BIBLIOGRAPHY

Here are the complete references to all publications cited in the text and to some other publications that present ideas and results closely tied to the subjects I have discussed.

Abrams, Robert. 1980. *Foundations of Political Analysis*. New York: Columbia University Press.

Aizerman, Mark A. 1983. "New Problems in the General Choice Theory." Mimeograph. Moscow: Institute of Control Sciences.

Aldrich, John. 1977. "The Dilemma of a Paretian Liberal: Some Consequences of Sen's Theorem." *Public Choice* 30:1–21.

Armstrong, W. E. 1939. "The Determinateness of the Utility Function." *Economic Journal* 49:453–467.

Arrow, Kenneth J. 1950. "A Difficulty in the Concept of Social Welfare." *Journal of Political Economy* 58:328–346.

_____ 1952. *Social Choice and Individual Values.* 1st ed. New York: Wiley.

_____ 1959. "Rational Choice Functions and Orderings." *Economica* [N.S.] 26:121–127.

_____ 1963. *Social Choice and Individual Values.* 2d ed. New Haven: Yale University Press. First edition published by Wiley, 1951.

_____ 1967a. "Public and Private Values." In Sidney Hook, ed., *Human Values and Economic Policy*, pp. 3–21. New York: New York University Press.

_____ 1967b. "The Place of Moral Obligation in Preference Systems." In Sidney Hook, ed., *Human Values and Economic Policy*, pp. 117–119. New York: New York University Press.

_____ 1977a. "Current Developments in the Theory of Social Choice." *Social Research* 44:607–622.

_____ 1977b. "Extended Sympathy and the Possibility of Social Choice." *American Economic Review* 67:219–225

d'Aspremont, Claude and Louis Gevers. 1977. "Equity and the Informational Basis of Collective Choice." *Review of Economic Studies* 44:199–209.

Axelrod, Robert. 1984. *The Evolution of Cooperation*, New York: Basic Books.

Banerji, D. 1964. "Choice and Order; or, First Things First." *Economica* [N.S.] 31:158–167.

Barry, Brian. 1965. *Political Argument*. New York: Humanities Press.

Batra, Raveendra N. and Prasanta Pattanaik. 1972. "On Some Suggestions for Having Non-Binary Social Choice Functions." *Theory and Decision* 3:1–11.

Bernholz, Peter. 1973. "Logrolling, Arrow Paradox and Cyclical Majorities." *Public Choice* 15:87–95.

_____ 1974a. "Is a Paretian Liberal Really Impossible?" *Public Choice* 20:99–108.

_____ 1974b. "Logrolling, Arrow Paradox and Decision Rules—a Generalization." *Kyklos* 27:49–61.

_____ 1975. "Logrolling and the Paradox of Voting." *American Political Science Review* 69:916–62.

_____ 1983. "A General Constitutional Possibility Theorem." Universitat Basel, Discussion Paper No. 76.

Black, Duncan. 1948. "On the Rationale of Group Decision-Making." *Journal of Political Economy* 56:23–24.

_____ 1958. *Theory of Committees and Elections.* Cambridge: Cambridge University Press.

_____ 1969. "On Arrow's Impossibility Theorem." *Journal of Law and Economics* 12:227–248.

Black, Duncan and R. A. Newing. 1951. *Committee Decisions with Complementary Valuation*. London: William Hodge.

Blair, Douglas H. 1975. "Path-Independent Social Choice Functions: A Further Result." *Econometrica* 43:173–174.

Blair, Douglas, Georges Bordes, Jerry S. Kelley, and Kotaro Suzumura. 1976. "Impossibility Theorems without Collective Rationality." *Journal of Economic Theory* 13:361–379.

Blair, Douglas and Robert Pollack. 1979. "Collective Rationality and Dictatorship: The Scope of the Arrow Theorem." *Journal of Economic Theory* 21: 186–194.

Blau, Julian H. 1957. "The Existence of Social Welfare Functions." *Econometrica* 25:302–313.

_____ 1971. "Arrow's Theorem with Weak Independence." *Economica* [N.S.] 38:413–420.

_____ 1972. "A Direct Proof of Arrow's Theorem." *Econometrica* 40:61–67.

_____ 1975. "Liberal Values and Independence." *Review of Economic Studies* 42:395–403.

_____ 1979. "Semiorders and Collective Choice." *Journal of Economic Theory* 21:195–206.

Blydenberg, John C. 1971. "The Closed Rule and the Paradox of Voting." *Journal of Politics* 33:57–71.

Borda, Jean-Charles. 1781. "Memoires sue les Elections au Scrutin." *Histoires de l'Academie Royale des Sciences.* English translation by A. de Grazia. "Mathematical Derivation of an Election System." *Isis* 44 (1953).

Bordes, Georges. 1976. "Consistency, Rationality, and Collective Choice." *Review of Economic Studies* 43:451–457.

Brams, Steven J. and Peter C. Fishburn. 1978. "Approval Voting." *American Political Science Review* 72:831–847.

Brandt, Richard B. 1967. "Personal Values and the Justification of Institutions." In Sidney Hook, ed., *Human Values and Economic Policy.* New York: New York University Press.

Braybrooke, David. 1968. "Let Needs Diminish that Preferences May Prosper." *Studies in Moral Philosophy*. American Philosophical Quarterly Monograph Series. Oxford: Basil Blackwell.

Brown, Donald J. 1975. "Aggregation of Preferences." *Quarterly Journal of Economics* 89:456–469.

Buchanan, James M. 1954. "Individual Choice in Voting and the Market." *Journal of Political Economy* 62:334–343.

Buchanan, James and Gordon Tullock. 1962. *The Calculus of Consent.* Ann Arbor: University of Michigan Press.

Campbell, Donald E. 1973. "Social Choice and Intensity of Preference." *Journal of Political Economy* 81:211–218.

_____ 1976. "Democratic Preference Functions." *Journal of Economic Theory* 12:259–272.

______ 1977. "Computational Criteria for Voting Systems." *British Journal of Political Science* 7:85–98.

Chernoff, Herman. 1954. "Rational Selection of Decision Functions." *Econometrica* 22:422–443.

Churchman, C. West. 1966. "On the Intercomparison of Utilities." In S. R. Krupp, ed., *The Structure of Economic Science: Essays on Methodology*, pp. 243–256. Englewood Cliffs, N.J.: Prentice-Hall.

Cohen, Linda. 1979. "Cyclic Sets in Multidimensional Voting Models." *Journal of Economic Theory* 21:1–12.

Coleman, James S. 1966. "The Possibility of a Social Welfare Function." *American Economic Review* 56:1105–1122.

Condorcet, the Marquis de. 1785. *Essai sur l'Application de l'Analyse a la Probabilite des Decisions Rendues a la Pluralite des Voix.* Paris.

Copeland, A. H. 1951. "A 'Reasonable' Social Welfare Function." Mimeo. University of Michigan Seminar on Applications of Mathematics to the Social Sciences, Ann Arbor.

Davidson, Donald, J. C. C. McKinsey, and Patrick Suppes. 1955. "Outline of a Formal Theory of Value, I." *Philosophy of Science* 22:140–160.

Davis, Otto A., Morris H. DeGroot, and Melvin J. Hinich. 1972. "Social Preference Orderings and Majority Rule." *Econometrica* 40:147–157.

Davis, Otto A. and Melvin J. Hinich. 1966. "A Mathematical Model of Policy Formation in a Democratic Society." In Joseph Bernd, ed., *Mathematical Applications in Political Science,* 2:175–208. Dallas: Southern Methodist University Press.

Deb, Rajat. 1977. "On Schwartz's Rule." *Journal of Economic Theory* 16:103–110.

______ 1983. "Binariness and Rational Choice." *Mathematical Social Science* 5:97–106.

Debreu, Gerard. 1954. "Representation of a Preference Ordering by a Numerical Function." In R. M. Thrall, D. H. Coombs, and R. L. Davis, eds., *Decision Processes*. New York: Wiley.

______ 1959. *Theory of Value*. New Haven: Yale University Press.

DeMeyer, F. and Charles Plott. 1970. "The Probability of a Cyclical Majority." *Econometrica* 38:345–354.

Dobra, John L. 1983. "An Approach to Empirical Studies of Voting Paradoxes: An Update and Extension." *Public Choice* 41:241–250.

Dodgson, C. L. (Lewis Carroll). 1876. *A Method of Taking Votes on More than Two Issues.* Oxford. Reprinted in Black (1958).

______ 1873. *A Discussion of the Various Methods of Procedure in Conducting Elections.* Oxford.

______ 1874. *Suggestions as to the Best Method of Taking Votes When More Than Two Issues Are to Be Voted On.* Oxford.

Downs, Anthony. 1957. *An Economic Theory of Democracy.* New York: Harper and Row.

______ 1961. "In Defense of Majority Voting." *Journal of Political Economy* 69:192–199. Reprinted in Thomas Schwartz, ed., *Freedom and Authority: An Introduction to Social and Political Philosophy* pp. 304–313. Encino, Calif.: Dickinson, 1973.

Enelow, James M. and Melvin Hinich. 1984. *An Introduction to the Spatial Theory of Voting.* Cambridge: Cambridge University Press.

Enelow, James M. and David H. Koehler. 1979. "Vote Trading in a Legislative Context: An Analysis of Cooperative and Noncooperative Strategic Voting." *Public Choice* 34:157–175.

Farquharson, Robin. 1969. *Theory of Voting.* New Haven: Yale University Press.

Ferejohn, John A. and David Grether. 1974. "On a Class of Rational Social Decision Procedures." *Journal of Economic Theory* 8:471–482.

______ 1977. "Weak Path Independence." *Journal of Economic Theory* 14:19–31.

Fine, Ben J. 1975. "Individual Liberalism in a Paretian Society." *Journal of Political Economy* 83:1277–1281.

Fine, Kit. 1973. "Conditions for the Existence of Cycles Under Majority and Non-Minority Rule." *Econometrica* 41:889–899.

Fishburn, Peter C. 1970a. "Arrow's Impossibility Theorem: Concise Proof and Infinite Voters." *Journal of Economic Theory* 2:103–106.

______ 1970b. "Condition for Simple Majority Decision Functions with Intransitive Individual Indifference." *Journal of Economic Theory* 2:354–367.

______ 1970c. "Intransitive Indifference in Preference Theory: A Survey." *Operations Research* 18:.207–228.

______ 1970d. "Intransitive Indifference with Unequal Indifference Intervals." *Journal of Mathematical Psychology* 7:144–149.

______ 1970e. "The Irrationality of Transitivity in Social Choice." *Behavioral Science* 15:119–123.

______ 1971a. "A Comparative Analysis of Group Decision Methods." *Behavioral Science* 16:538–644.

______ 1971. "Should Social Choice Be Based on Binary Comparisons?" *Journal of Mathematical Sociology* 1:133–142.

______ 1972. *Mathematics of Decision Theory.* The Hague: Mouton.

______ 1973. *The Theory of Social Choice.* Princeton, N.J.: Princeton University Press.

______ 1973. "Transitive Binary Social Choices and Intraprofile Conditions." *Econometrica* 41:603–615.

______ 1974a. "Choice functions on Finite Sets." *International Economic Review* 15:729–749.

______ 1974b. "On Collective Rationality and a Generalized Impossibility Theorem." *Review of Economic Studies* 41:445–459.

______ 1974a. "Paradoxes of Voting." *American Political Science Review* 69:943–946.

______ 1974c. "Social Choice Functions." *SIAM Review* 16:63–90.

______1974b. "Subset Choice Conditions and the Computation of Social Choice Sets." *Quarterly Journal of Economics* 88: 320–329.

______ 1975. "Semiorders and Choice Functions." *Econometrica* 43: 975–977.

______ 1976. "Dictators on Blocks: Generalizations of Social Choice Impossibility Theorems." *Journal of Combinatorial Theory B* 20:153–170.

______ 1977. "Condorcet Social Choice Functions." *SIAM Journal of Applied Mathematics* 33:469–489.

Frohlich, Norman and Joe A. Oppenheimer. 1978. *Modern Political Economy*. Englewood Cliffs, N.J.: Prentice-Hall.

Galston, William A. 1980. *Justice and the Human Good.* Chicago: University of Chicago Press.

Gardenfors, Peter. 1973. "Positionalist Voting Functions." *Theory and Decision* 4:1–24.

Garman, Mark and Morton Kamien. 1968. "The Paradox of Voting: Probability Calculations." *Behavioral Science* 13:306–316.

Gorgescu-Roegen, Nicholas. 1969. "The Relation Between Binary and Multiple Choices: Some Comments and Further Results." *Econometrica* 37:728–730.

Gibbard, Allan. 1973. "Manipulation of Voting Schemes: A General Result." *Econometrica* 41:587–601.

_____ 1974. "A Pareto Consistent Libertarian Claim." *Journal of Economic Theory* 7:388–410.

_____ 1977. "Manipulation of Schemes that Mix Voting with Chance." *Econometrica* 45:665–681.

Goodman, L.A. and H. Markowitz. 1952. "Social Welfare Functions Based on Individual Rankings." *American Journal of Sociology* 58:257–262.

Hansson, Bengt. 1968. "Choice Structures and Preference Relations." *Synthese* 18:443–458.

_____ 1969. "Voting and Group Decision Functions.." *Synthese* 20:526–537.

_____ 1973. "The Independence Condition in the Theory of Social Choice." *Theory and Decision* 4:25–49.

Harsanyi, John C. 1955. "Cardinal Welfare, Individualistic Ethics, and Interpersonal Comparisons of Utility." *Journal of Political Economy* 63:309–321.

Herzberger, Hans G. 1973. "Ordinal Preference and Rational Choice." *Econometrica* 41:187–237.

Hotelling, Harold. 1929. "Stability in Competition." *The Economic Journal* 39:41–57. Reprinted in George J. Stigler and Kenneth E. Boulding, eds., *Readings in Price Theory,* pp. 467–485. Chicago: Irwin, 1952.

Houthakker, Hendrik S. 1950. "Revealed Preference and the Utility Function." *Economica* 17: 159–174.

_____ 1965. "On the Logic of Preference and Choice." In A. T. Tyminieniecka, ed., *Contributions to Logic and Methodology,* pp. 193–207. Amsterdam: North Holland.

Howard, Nigel. 1971. *Paradoxes of Rationality: Theory of Metagames and Political Behavior*. Cambridge, Mass.: M.I.T. Press.

Inada, K.-I. 1969. "On the Simple Majority Decision Rule." *Econometrica* 37:490–506.

_____ 1970. "Majority Rule and Rationality." *Journal of Economic Theory* 2:27–40.

Jamison, Dean T. and Lawrence J. Lau. 1973. "Semiorders and the Theory of Choice." *Econometrica* 41:901–912.

_____ 1975. "Semiorders and the Theory of Choice: A Correction." *Econometrica* 43:975–977.

Kadane, J. B. 1972. "On Division of the Question." *Public Choice* 13:47–54.

Kalai, Ehud, Eliska Panzer, and David Schmeidler. 1976. "Collective Choice Correspondences as Admissible Outcomes of Social Bargaining Processes." *Econometrica* 44:233–240.

Kalai, Ehud and David Schmeidler. 1977. "Aggregation Procedure for Cardinal Preferences: A Formulation and Proof of Samuelson's Impossibility Conjecture." *Econometrica* 45:431–438.

Karni, Edi and David Schmeidler. 1976. "Independence of Non-Feasible Alternatives, and Independence of Non-Optimal Alternatives." *Journal of Economic Theory* 12:488–493.

Kelley, Jerry S. 1976a. "The Impossibility of a Just Liberal." *Economica* 43:67–75.

______ 1976b. "Rights-Exercising and a Pareto Consistent Libertarian Claim." *Journal of Economic Theory* 13:138–153.

______ 1977. "Strategy-Proofness and Social Choice Functions without Single Valuedness." *Econometrica* 45:439–446.

______ 1978. *Arrow Impossibility Theorems.* New York: Academic Press.

Kirman, Alan P. and Dieter Sonderman. 1972. "Arrow's Theorem, Many Agents, and Invisible Dictators." *Journal of Economic Theory* 5:267–277.

Koehler, David H. 1975. "Vote Trading and the Voting Paradox: A Proof of Logical Equivalence." *American Political Science Review* 69:954–960.

Kramer, Gerald. 1973. "On a Class of Equilibrium Conditions for Majority Rule." *Econometrica* 41:285–297.

Larson, Bo. 1983. *Basic Properties of Majority Rule.* Lund, Sweden: Department of Economics, University of Lund.

Little, I.M.D. 1952. "Social Choice and Individual Values." *Journal of Political Economy* 60:422–432.

______ 1957. *A Critique of Welfare Economics.* 2d ed. (1st ed. 1950). Oxford: Clarendon Press.

Luce, R. Duncan. 1956. "Semi-Orders and a Theory of Utility Discrimination." *Econometrica* 24:178–191.

Luce, R. Duncan and Howard Raiffa. 1957. *Games and Decisions.* New York: Wiley.

McCubbins, Mathew and Thomas Schwartz. 1985. "The Politics of Flatland." *Public Choice* 46:45–60.

McGarvey, D. C. 1953. "A Theorem on the Construction of Voting Paradoxes." *Econometrica* 21:608–610.

MacKay, Alfred F. 1975. "Interpersonal Comparisons." *Journal of Philosophy* 72:535–549.

_____ 1980. *Arrow's Theorem: The Paradox of Social Choice.* New Haven: Yale University Press.

McKelvey, Richard D. 1976. "Intransitivities in Multidimensional Voting Models, and Some Implications for Agenda Control." *Journal of Economic Theory* 2:472–482.

Margolis, Julius. 1955. "A Comment on the Pure Theory of Public Expenditure." *Review of Economics and Statistics* 35:347–349.

Mas-Colell, Andreu and Hugo F. Sonnenschein. 1972. "General Possibility Theorems for Group Decisions." *Review of Economic Studies* 39:185–192.

May, K. O. 1952. "A Set of Independent, Necessary, and Sufficient Conditions for Simple Majority Decision." *Econometrica* 20:680–684.

_____ 1953. "A Note on Complete Independence of the Conditions for Simple Majority Decision." *Econometrica* 21:172–173.

_____ 1954. "Intransitivity, Utility and the Aggregation of Preference Patterns." *Econometrica* 22:1–13.

May, Robert M. 1971. "Some Mathematical Remarks on the Paradox of Voting." *Behaviorall Science* 16:143–51.

Mayston, David. 1975. *The Idea of Social Choice.* New York: St. Martin's Press.

Miller, Nicholas. 1975. "Logrolling and the Arrow Paradox: A Note." *Public Choice* 21:107–110.

_____ 1977a. "Graph Theoretical Approaches to the Theory of Voting." *American Journal of Political Science* 21:769–803.

_____ 1977b. "Logrolling, Vote Trading, and the Paradox of Voting: A Game Theoretical Overview." *Public Choice* 30:51–73.

_____ 1980. "A New Solution Set for Tournaments and Majority Voting: Further Graph Theoretical Approaches to the Theory of Voting." *American Journal of Political Science* 24:68–96.

Mueller, Dennis. 1979. *Public Choice.* Cambridge: Cambridge University Press.

Murakami, Yasusuke. 1961a. "A Note on the General Possibility Theorem of the Social Welfare Function." *Econometrica* 29:244–246.

_____ 1961b. "Some Logical Properties of Arrovian Social Welfare Function." *The Journal of Economic Behavior* 1:77–84.

_____ 1968. *Logic and Social Choice.* New York: Dover.

Nanson, E. J. 1882. "Methods of Elections." *Transactions and Proceedings of the Royal Society of Victoria* 18:197–240.

Neuman, John von and Oskar Morgenstern. 1953. *Theory of Games and Economic Behavior.* 3d ed. New York: Wiley.

Niemi, Richard G. and Herbert Weisberg. 1968. "A Mathematical Solution for the Probability of the Paradox of Voting." *Behavioral Sciences* 13:317–323.

Olson, Mancur. 1965. *The Logic of Collective Action.* Cambridge, Mass.: Harvard University Press.

Oppenheimer, Joe A. 1972. "Relating Coalitions of Minorities to the Voter's Paradox, or Putting the Fly in the Democratic Pie." Paper presented to meeting of the Southwest Political Science Association, San Antonio, March.

______ 1975. "Some Political Implications of 'Vote Trading and the Voting Paradox: A Proof of Logical Equivalence.'" *American Political Science Review* 69:963–966.

Osborne, D. K. 1975. "On Liberalism and the Pareto Principle." *Journal of Political Economy* 83:1283–1287.

______ 1976. "Irrelevant Alternatives and Social Welfare." *Econometrica* 44:1001–1015.

Packard, Dennis J. 1975a. "A Note on Wittgenstein and Cyclical Comparatives." *Analysis* 36:37–40.

______ 1975b. "On the Impossibility of Measurement." *Utah Academy Proceedings* 52 (part 2):36–43.

______ 1975c. "Social Choice Theory and Citizens' Intransitive Weak Preference." *Public Choice* 22:107–111.

______ 1982. "Cyclical Preference Logic." *Theory and Decision* 14: 415–426.

Page, Benjamin. 1977. "Elections and Social Choice: The State of the Evidence." *American Journal of Political Science* 21:639–668.

Parks, Robert P. 1976a. "Further Results on Path Independence, Quasi-Transitivity and Social Choice." *Public Choice* 26:75–87.

______ 1976b. "An Impossibililty Theorem for Fixed Preferences: A Dictatorial Bergson-Samuelson Welfare Function." *Review of Economic Studies* 43:447–450.

Pattanaik, Prasanta K. 1971. *Voting and Collective Choice.* London and New York: Cambridge University Press.

______ 1978. *Strategy and Group Choice.* Amsterdam: North Holland.

Plott, Charles. 1967. "A Notion of Equilibrium and Its Possibility Under Majority Rule. *American Economic Review* 57:788–806.

______ 1971. "Recent Results on the Theory of Voting." In Michael Intrilligator, ed., *Frontiers in Quantitative Economics.* Amsterdam: North Holland.

______ 1972a. "Ethics, Social Choice Theory, and the Theory of Economic Policy." *Journal of Mathematical Sociology* 2:181–208.

______ 1972b. "Individual Choice of a Decision Process." In Richard Niemi and Herbert Weisberg, ed., *Probability Models of Collective Decision Making,* pp. 83–87. Columbus, Ohio: Bobbs-Merrill.

______ 1973. "Path Independence, Rationality, and Social Choice." *Econometrica* 41:1075–1091.

______ 1976. "Axiomatic Social Choice Theory: An Overview and Interpretation." *American Journal of Political Science* 20:511–596.

Ray, Paramesh. 1973. "Independence of Irrelevant Alternatives." *Econometrica* 41:987–991.

Rescher, Nicholas. 1969. *Introduction to Value Theory.* Englewood Cliffs: N.J.: Prentice-Hall.

______ 1972. "Some Observations on Consensus Methodology." *Theory and Decision* 3:175–179.

Richelson, Jeffrey T. 1975. "A Comparative Analysis of Social Choice Functions." *Behavioral Science* 20:331–337.

______ 1978a. "A Comparative Analysis of Social Choice Functions, II." *Behavioral Science* 23:38–44.

______ 1978b. "A Comparative Analysis of Social Choice Functions, III." *Behavioral Science* 23:169–176.

______ 1978c. "Some Further Results on Consistency, Rationality, and Collective Choice." *Review of Economic Studies* 45:343–346.

______ 1978d. "A Characterization Theorem for the Plurality Rule." *Journal of Economic Theory* 19:348–350.

______ 1981. "Majority Rule and Collective Choice." Manuscript.

Richter, Marcel R. 1966. "Revealed Preference Theory." *Econometrica* 34:635–645.

______ 1971. "Rational Choice." In J. S. Chipman, L. Hurwicz, M. K. Richter, and H. F. Sonnenschein, eds., *Preferences, Utility, and Demand,* pp. 29–58. New York: Harcourt Brace Jovanovich.

Riker, William H. 1958. "The Paradox of Voting and Congressional Rules for Voting on Amendments." *American Political Science Review* 52:349–366.

______ 1961. "Voting and the Summation of Preferences: An Interpretive Bibliographic Review of Selected Developments During the Last Decade." *American Political Science Review* 55:900–911.

______ 1980. "Implications from the Disequilibrium of Majority Rule for the Study of Institutions." *American Political Science Review* 74: 432–458. Reprinted in Peter C. Ordeshook and Kenneth A. Shepsle, eds., *Political Equilibrium*, pp. 3–24. Boston: Kluwer-Nijhoff, 1982.

______ 1982. *Liberalism Against Populism.* San Francisco: Freeman.

Riker, William H. and Steven Brams. 1973. "The Paradox of Vote Trading." *American Political Science Review* 67:1235–1247.

Riker, William H. and Peter Ordeshook. 1973. *An Introduction to Positive Political Theory.* Englewood Cliffs, N.J.: Prentice-Hall.

Rothenberg, Jerome. 1961. *The Measurement of Social Welfare.* Englewood Cliffs, N.J.: Prentice-Hall.

Salles, Maurice. 1975. "A General Possibility Theorem for Group Decision Rules with Pareto Transitivity." *Journal of Economic Theory* 2:110–118.

Samuelson, Paul A. 1938. "A Note on the Pure Theory of Consumer's Behavior." *Economica* 5:61–71.

______ 1954. "The Pure Theory of Public Expenditures." *Review of Economics and Statistics* 36:347–349, 387, 389.

______ 1967. "Arrow's Mathematical Politics." In Sidney Hood, ed., *Human Values and Economic Policy*, pp. 41–52. New York: New York University Press.

______ 1972. "Maximum Principles in Analytical Economics." *American Economic Review* 62:249–262.

Saposnik, Rubin. 1975. "On the Transitivity of the Social Preference Relation under Simple Majority Rule." *Journal of Economic Theory* 10:1–7.

Satterthwaite, Mark A. 1975. "Strategy-Proofness and Arrow's Conditions: Existence and Correspondence Theorems for Voting Procedures and Social Welfare Functions." *Journal of Economic Theory* 10:187–217.

Schofield, Norman. 1978. "Instability of Simple Dynamic Games." *Review of Economic Studies* 45:575–594.

Schwartz, Thomas. 1970. "On the Possibility of Rational Policy Evaluation." *Theory and Decision* 1:89–106.

_____ 1972. "Rationality and the Myth of the Maximum." *Noûs* 6:97–117.

_____ 1975a. "On the Utility of MacKay's Comparisons." *Journal of Philosophy* 72:549–551.

_____ 1975b. "Serial Collective Choice." Manuscript. Pittsburgh: Carnegie-Mellon University.

_____ 1975c. "Vote Trading and Pareto Efficiency." *Public Choice* 24: 101–109.

_____ 1976. "Choice Functions, 'Rationality' Conditions, and Variations on the Weak Axiom of Revealed Preference." *Journal of Economic Theory* 13:414–427.

_____ 1977. "Collective Choice, Separation of Issues, and Vote Trading." *American Political Science Review* 71:999–1010.

_____ 1981. "The Universal Instability Theorem." *Public Choice* 37:487–501.

_____ 1982a. "Human Welfare: What It Is Not." In Harlan Miller and William Williams, eds., *The Limits of Utilitarianism*, pp. 195–206. Minneapolis: University of Minnesota Press.

_____ 1982b. "No Minimally Reasonable Collective Choice Process Can Be Strategy-Proof." *Mathematical Social Sciences* 3:57–72.

_____ 1982c. "A Really General Impossibility Theorem." *Quality and Quantity* 16:493–505.

_____ 1985. "Dr. Krankheit and the Concept of Compliance." *American Philosophical Quarterly* 22:81–87.

Schwartz, Thomas, ed. 1973. *Freedom and Authority: An Introduction to Social and Political Philosophy.* Encino, Calif.: Dickinson.

Schick, Frederick. 1969. "Arrow's Proof and the Logic of Preference." *Philosophy of Science* 36:127–144.

Sen, Amartya. 1966. "A Possibility Theorem on Majority Decisions." *Econometrica* 34:491–499.

_____ 1970a. *Collective Choice and Social Welfare.* San Francisco: Holden-Day.

_____ 1970b. "The Impossibility of a Paretian Liberal." *Journal of Political Economy* 78:152–157.

_____ 1970c. "Interpersonal Aggregation and Partial Comparability." *Econometrica* 38:393–409.

______ 1971. "Choice Functions and Revealed Preference." *Review of Economic Studies* 38:307–317.

______ 1973. "Behavior and the Concept of Preference." *Economica* [N.S.] 40:241–259.

______ 1976. "Liberty, Unanimity, and Rights." *Economica* 43: 217–245.

______ 1977. "Social Choice Theory: A Re-Examination." *Econometrica* 45:53–89.

______ 1983. "Social Choice Theory." In Kenneth J. Arrow and Michael D. Intriligator, eds., *Handbook of Mathematical Economics,* vol. 3. Amsterdam: North Holland.

Sen, Amartya K. and Prasanta Pattanaik. 1969. "Necessary and Sufficient Conditions for Rational Choice Under Majority Decision." *Journal of Economic Theory* 1:178–202.

Shepsle, Kenneth A. 1979. "Institutional Arrangements and Equilibrium in Multidimensional Voting Models." *American Journal of Political Science* 23:27–59.

Shepsle, Kenneth A. and Barry Weingast. 1981. "Structure-Induced Equilibrium and Legislative Choice." *Public Choice* 37:503–519.

Smith, John. 1073. "Aggregation of Preferences with Variable Electorate." *Econometrica* 41:1027–1041.

Strasnick, Stephen. 1976. "The Problem of Social Choice: Arrow to Rawls." *Philosophy and Public Affairs* 5:241–273.

Sullivan, Terry. 1976. "Voter's Paradox and Logrolling: An Initial Framework for Committee Behavior on Appropriations." *Public Choice* 25:31–45.

Suppes, Patrick. 1960. *Axiomatic Set Theory.* Princeton, N.J.: Van Nostrand.

Suzumura, Kotaro. 1976. "Remarks on the Theory of Collective Choice." *Economica* 43:381–390.

______ 1983. *Rational Choice, Collective Decisions, and Social Welfare.* Cambridge: Cambridge University Press.

Taylor, Michael. 1976. *Anarchy and Cooperation.* New York: Wiley.

Tullock, Gordon. 1959. "Problems of Majority Voting." *Journal of Political Economy* 67:571–579. Reprinted in Thomas Schwartz, ed., *Freedom and Authority: An Introduction to Social and Political Philosophy*, pp. 295–303. Encino, Calif.: Dickinson, 1973.

______ 1964. "The Irrationality of Intransitivity." *Oxford Economic Papers* [N.S.] 16:401–406.

______ 1967a. "The General Irrelevance of the General Impossibility Theorem." *Quarterly Journal of Economics* 81:256–270.

______ 1967b. *Toward a Mathematics of Politics.* Ann Arbor: University of Michigan Press.

______ 1981. "Why So Much Stability?" *Public Choice* 37:189–204.

Tversky, Amos. 1969. "Intransitivity of Preferences." *Psychological Review* 76:31–48.

Vickrey, W. 1960. "Utility, Strategy, and Social Decision Rules." *Quarterly Journal of Economics* 74:507–535.

Waldner, Ilmar. 1972. "The Empirical Meaningfulness of Interpersonal Utility Comparisons." *Journal of Philosophy* 69:87–103.

______ 1973. "The Possibility of Rational Policy Evaluation." *Theory and Decision* 4:85–90.

Ward, Benjamin. 1960. "Majority Rule and Allocation." *Journal of Conflict Resolution* 5:380–389.

______ 1965. "Majority Voting and Alternative Forms of Public Enterprises." In Julius Margolis, ed., *The Public Economy of Urban Expenditures.* Baltimore: Johns Hopkins University Press.

Wilson, Robert B. 1969. "An Axiomatic Model of Logrolling." *American Economic Review* 59:331–341.

______ 1970. "The Finer Structure of Revealed Preference." *Journal of Economic Theory* 2:348–353.

______ 1972a. "The Game-Theoretic Structure of Arrow's General Possibility Theorem." *Journal of Economic Theory* 5:14–20.

______ 1972b. "Social Choice Theory Without the Pareto Principle." *Journal of Economic Theory* 5:478–486.

______ 1975. "On the Theory of Aggregation." *Journal of Economic Theory* 10:89–99.

Young, H. P. 1974. "An Axiomatization of Borda's Rule." *Journal of Economic Theory* 9:43–52.

______ 1975. "Social Choice Scoring Functions." *SIAM Journal of Applied Mathematics.* 28:824–838.

Zeckhauser, Richard. 1969. "Majority Rule with Lotteries on Alternatives." *Quarterly Journal of Economics* 83:696–703.

INDEX